THE KEY
TO THE PROBLEMS OF EXISTENCE

The Initiatic Teaching of the Universal White Brotherhood

Omraam Mikhaël Aïvanhov

THE KEY
TO THE PROBLEMS OF EXISTENCE

Translated from the French
New Edition

Volume 11
from the Complete Works

ÉDITIONS PROSVETA

Distributed by:

AUSTRIA — HELMUTH FELDER VERLAG
Kranebitterallee 88/144 – Postfach 33
A-6027 Innsbruck

BELGIUM — VANDER S.A. – Av. des Volontaires 321
B - 1150 Bruxelles

BRITISH ISLES — PROSVETA Ltd. – 4 St. Helena Terrace
Richmond, Surrey TW9 1NR
Trade orders to :
ELEMENT Books Ltd
Unit 25 Longmead Shaftesbury
Dorset SP7 8PL

CANADA — PROSVETA Inc. – 1565 Montée Masson
Duvernay est, Laval, Qué. H7E 4P2

DENMARK — SANKT ANSGARS FORLAG
Bredgade 67
DK – 1260 COPENHAGUE

FRANCE — Editions PROSVETA S.A. – B.P. 12
83601 Fréjus Cedex

GERMANY — URANIA – Rudolf Diesel Ring 26
D - 8029 Sauerlach

GREECE — PROSVETA HELLAS
90 Bd. Iroon Polytechniou
185 36 Le Pirée

HOLLAND — MAKLU B.V. – Koninginnelaan 96
NL - 7315 EB Apeldoorn

HONG KONG — HELIOS
31 New Kap Bin Long Village
Sai Kung N.T., Hong Kong

IRELAND — PROSVETA IRELAND
24 Bompton Green
Castleknock, Dublin

ITALY — PROSVETA – Bastelli 7
I - 43036 Fidenza (Parma)

PORTUGAL — Edições IDADE D'OURO
Rua Passos Manuel 20 – 3.° Esq.
P - 1100 Lisboa

SPAIN — PROSVETA ESPAÑOLA – Caspe 41
Barcelona – 10

SWITZERLAND — PROSVETA Société Coopérative
CH - 1801 Les Monts-de-Corsier

UNITED-STATES — PROSVETA U.S.A. – P.O. Box 49614
Los Angeles, California 90049

Enquiries should be addressed to the nearest distributor

Édition originale ISBN 2-85566-231-1
ISBN 2-85566-313-X

TABLE OF CONTENTS

OMRAAM MIKHAËL AÏVANHOV

Foreword

The whole world talks about change, the whole world wants change, yet nothing ever changes. Why? Because man needs a projector such as the Initiatic Science to show him clearly what change actually is, and what must be done in order to bring it about.

Real change begins with the person who succeeds in installing peace, order and harmony within himself, dynamics that set in motion a continual, uninterrupted movement of perfectioning. Our inner world is a field of battle, a stamping ground for our conflicting tendencies: our generous impulses versus our selfishness, our disinterested thinking versus our self-interested calculations, our personal ambition versus the good of all. As soon as the mind chooses a certain course, the heart raises objections and vice versa, and our behaviour is erratic and contradictory because of this inner conflict. A coherent line of behaviour resulting from the harmonious link between thoughts, emotion and action, is to be found only in beings with exceptional strength of character, who are completely dedicated to a glorious cause. This alone gives the ability to dominate and control the inner world and, because of that control, bring about a real change in the external world. However, as soon as this highly evolved being disappears, if there is no one to carry on his way of thinking, the change he effected wears away little by little until it too disappears.

Only a man who has succeeded in improving himself can bring about the changes that will improve the world.

Man is distinguishable from other creatures by the fact that he is self-conscious, that is, he has the ability to consider, observe and analyse the inner and invisible but most active world of his thoughts, emotions and will. But the Self he is becoming conscious of can never be fully grasped once and for all, for he is still for the most part unconscious. We can bring to the surface of our consciousness only an infinitesimal part of the tremendous reserves of energy and imagery stored in our unconscious, which is why we are caught by surprise by our dreams, our sudden emotions, our unpredictable moods... all our obsessions that colour the psychic life and cristallize into behaviour. The unconscious self is actually our human nature, our character which is hard to control, and to which all are subject who have not yet embarked on the spiritual life, the real work.

Our human nature affects us in two different ways. One is explained by psychoanalysts who claim that any eruption on the part of the unconscious into the conscious, is so disturbing (because so out of line with what society expects of us) that the conscious is ashamed and tries to repress these eruptions, the childish narcissistic attitude that insists on being the sole object of someone's affections, the pride that competes with the Author of all things, the desire for revenge which would eliminate all who impede its search for pleasure. Under the pressure of educational and social standards, the conscious sets up a system of defense, say the psychoanalysts, that censures and represses all our instinctive urges, our primitive egocentricities... all the more so because it refuses to recognize in itself any such anti-social tendencies!

Psychoanalysts do not however go into the other expressions of the unconscious, which, despite their being uncon-

scious, we would not for anything in the world repress. The impulses of courage and hope that open up the world of harmony, for instance, the subtle joys at discovering celestial purity, the springing forth of creative light, the indestructible unity of mankind at the level of the soul and Spirit, the feeling of Immortality, of Eternity... all these brush lightly by our consciousness without it being able to seize them, in spite of our desire to be included in this new expansion of sensation and perception.

Consciousness is the mirror of Heaven and Hell, but it is powerless to create either. We have within us two natures, both unconscious, and it depends on the way we lead our daily lives, says Omraam Mikhaël Aïvanhov, whether we have experiences of light, beauty, balance, or discord, disorder, anguish, terror.

Now we must bring the other, unconscious nature, the higher Self (as compared to the first or lower nature) into the light. This is extremely important for our education, our psychology, our understanding of social problems. The terms lower and higher indicate the importance to attach to each one, where to situate them. Man achieved a physiological victory once he assumed a vertical position with his head above the level of his stomach and sex; now he must achieve a psychological victory by assuming an upright position within himself. He must be able to identify and situate within his inner hierarchy the egocentric impulses that urge him to seek gross satisfactions, limiting him, paralyzing him, as he must the vast and luminous inspirations from the higher nature that spur his mind and heart to devote his existence to the good of all.

These two natures, lower and higher, are called by Master Omraam Mikhaël Aïvanhov the personality and the individuality. The personality, egocentric and self-interested, versatile and unreliable, cruel or obsequious (whichever is to its advantage) thinks of everything in terms of its own pleasure and personal enjoyment. The Latin word "persona" from which *stems personality, is the mask worn in the theatre indicating disguise, deception, duplicity. The individuality stands for man's indivisible nature, the pure and simple essence of the Spirit without which he would not exist.*

The personality and the individuality have the same structure, that is, both are triads with three means of expression: thought, emotion, action. On the lower level of the personality, the mind is the seat of erroneous thinking, of ruse and craft, the heart is the seat of feelings of hate and possessiveness, and the will is the seat of violent and destructive action. In the higher triad, the mind is the centre of thoughts of universality, it knows the great laws of existence and uses them to light its path toward a life of brotherhood, of benefit to all; the soul is the centre of feelings of generosity, compassion, forgiveness, of worship and praise toward God; and the Spirit is the centre of creative action, of liberation from the shackles of the personality, it revives the inner world of all creatures and reanimates the feeling that all creatures belong to God.

By this identical triad formation, the Master Omraam Mikhaël Aïvanhov shows that our two natures are meant to correspond, and to be united. The power of the higher nature is at its greatest when it expresses itself through the lower nature's thoughts, feelings and actions, thus realizing the unsuspected power of the soul and Spirit. In the same way, if you want to reach God, you must mobilize all the energies of your lower nature, make the raw sap rise up the inner tree to feed the leaves, flowers and fruit of the higher nature at the top.

When the circulation is allowed to flow uninterruptedly between the lower and higher natures, it gives us the fullest use of all our faculties... that is when the great change takes place and we become like the sun! The sun shines on one and all alike with complete lack of prejudice, the sun warms all creatures with a love that is pure and disinterested, and the Spirit of the sun vivifies and creates all things. The sun, centre of our planetary system, origin of all life on earth, has been effecting endless transformations over millions of years, pushing mankind toward an ever increasingly complex order. The inner sun of our individuality is all-powerful as far as our earth, the personality, is concerned.

Do you see what an inexhaustible subject this is for anyone who wants to explore the two natures?

The first thing is to realize that most of us live with only half of our being, our personality, and go on looking amid the disorder, contradictions and limitations of the personality for the peace, love, happiness, success and intelligence we long for. We always can find a few crumbs of course, precisely because of the link that exists between the two natures, but now it is time for man to climb back up to the heights of his being, to the summit, where he will find all he desires in great force, abundance and perfection, and then to translate what he has found into behaviour.

How many discoveries we would make if we took this key, this knowledge of the two influences, the personality and the individuality, and inserted it into the door of the world of politics! Public speakers always use the words of the individuality (for the good of all, the rights of man and happiness for all, the rights of children, the liberation of the oppressed, aid to the underprivileged) but they themselves listen to the promptings of the personality: put personal interests above all, work less and make more by exploiting others, profit by pretending to help the helpless and then take over....

This discrepancy between speech and behaviour is the ideal trick of the trade of the lecturing politician, the anodyne that soothes the listener and hides the fact that the speaker is powerless to bring about any change whatsoever.

Hobbe's words, "dog eat dog", describing man's inhumanity to man, are reiterated continually with great success in political circles, because it confirms man's attitude, his way of thinking, feeling and acting on the level of his lower nature, the personality. All conflict comes from the different egocentric interests of two or more colliding personalities. It is easy to see how vitally important to the world it would be to have every nation, every statesman, every individual with a sense of responsibility, receive an education of universality so that all realize the need to enlarge their consciousness to the scale of all mankind. Their decisions would then correspond to the people's needs and desires. If a man's own tendencies were purified, under control and harmonized inside himself, he would shine forth with the light of his higher nature, and then peace, cohesion, unity would be the lot of those in his charge and all whom he serves. Everyone, world leader or no, must do this work in his immediate sphere, no one else can do it for him since his inner world is his alone... and that is where the kingdom must be won.

This idea that there is no alternative (if we are to preserve the integrity of both the individual and society) to acceptance of the existence of the higher Self, the individuality in everyone, shows us the problem of education and upbringing in a new light. Sociologists give as the reason for primitive societies being more cohesive than our modern society, the rites of passage, or initiation of the child-adolescent into adulthood, into the world of his elders, rites that have disappeared with the coming of industrialization. Today the adolescent is left to face his transition alone, and for him it is a period of crisis, of revolt and sometimes violent acts of aggression against the society he despises and wants no part of, and even against himself in his isolation.

The Master Omraam Mikhaël Aïvanhov makes it clear that it is not a question of going back into the past to find a more solid structure for society. The rites of passage, the transition from childhood to manhood, took place as part of magic ceremonies, with trial by fire, tattooing of candidates and so on, rites which are not possible today, our contemporary way of thinking would not allow it. But the essential must be retained, the Initiatic meaning, which is simply the transition from the personality to the individuality. Childish narcissism is taken for granted in the child since that is what makes him develop, but when he becomes an adult, he must set aside egocentric behaviour.

The need to surpass himself felt by the adolescent is not understood by adults who are immature and still caught up in the contradictions of their lower nature. Adolescence is then a time of emptiness, haunted by unrealizable dreams and filled with resentment against the society the youth rejects but must nevertheless accept since he has nothing better to offer! This emptiness, this void is accompanied by a more or less acute feeling of death. Now, it is necessary to die in order to live; that is the profound meaning of Initiation. To die as far as the personality-prison is concerned, in order to live in the individuality-freedom! But how can an adult who has not begun this inner metamorphosis be of any help to an adolescent? It is a change that alone brings progress and perfection and... change to the world. Freedom cannot exist outwardly until man has learned to control his inner impulses and is free of their influence; peace cannot exist in the world until man is at peace with himself; harmony cannot exist on earth until man is in harmony with himself, all the way to the vibrations in the cells of his physical body... then and then only can he spread harmony around him.

If that be so, then it is our understanding of these two natures that could solve our problems and eventually change mankind. We would dedicate all our learning and technology to solving problems on a planetary scale, our love would be obliged to reach further than the egocentric love of family and state, and we would perceive at last the spark of Divinity in each other! War could not exist under those conditions, the individuality in its superconsciousness cannot but feel the unity that exists between men. If all men were to feel the evil they are thinking of inflicting on others, they would not do it.

As long as we maintain and spread the philosophy of the

VIII

personality, whose point of reference is the body, matter, separateness, isolation, we will keep trying to solve man's problems by treating him as if he were a piece of flesh to be tossed aside, or cut up, or tortured, or killed. Yes, we may think the problem is solved that way, but only in appearance and only momentarily. For the immortal soul can bring other souls to bear, and Karma is ever present to extract payment for our deeds. Any change thus obtained is merely part of the infernal cycle of revenge... nothing changes, everything repeats itself indefinitely.

The Master Omraam Mikhaël Aïvanhov does not advocate unilateral disarmament in any form. The solution lies in ourselves, the work must be done by all men on themselves if ever the individuality is to gain the upper hand. And no one would be out of work, no unemployment! If the leaders of every country adopted these ideas, it would bring about changes that would be felt throughout society on every level, for the greatest good of all and for the survival of the species.

A golden key has been put into our hands by Master Omraam Mikhaël Aïvanhov, which is indeed the solution to all our problems, but we must be under no illusion: the stakes are high and we stand to gain tremendously, but the work is long and arduous. Will one life be enough for us to bring these two natures into harmony, enough to make visible the invisible light of the Spirit?

The reader cannot fail to be deeply stirred by the abundant images and examples and explanations this book gives, crisscrossing the map of our existence from left to right, from above to below, from the centre to the periphery. As with two competing teams, each player must be followed in order to see the whole game. Twenty-five lectures are hardly enough to teach us the rules of combat between our lower and higher natures, or how to gain a definite victory in favour of the real creator of change, our higher Self. Repetition is part of the festive atmosphere, part of the suspense and enthusiasm always present at great tourneys. In spite of the practically identical positions of *the players, the issue is eternally uncertain, for life, the great improvisor, is ever on our heels, spurring us on by presenting us ever greater problems to solve, requiring ever more intelligence, more love and more control, if what we want is change. The greatest change of all would be the realization of the Universal White Brotherhood.*

Le Bonfin, 17 July, 1983

Agnes Lejbowicz, University Professor
(translated from the French)

EDITOR'S NOTE

The reader is asked to bear in mind that the Editors have retained the spoken style of the Maître Omraam Mikhaël Aïvanhov in his presentation of the Teaching of the great Universal White Brotherhood, the essence of the Teaching being the spoken Word.

They also wish to clarify the point that the word *white* in Universal White Brotherhood, does not refer to colour or race, but to purity of soul. The Teaching shows how all men without exception (universal), can live a new form of life on earth (brotherhood), in harmony (white), and with respect for each other's race, creed and country... that is, Universal White Brotherhood.

Chapter 1

The Personality

Question: Master, you say that the personality is not divine by nature but how do you explain that, since nothing exists outside of God?

The question is an important one, but difficult to answer. The word "divine" has two meanings. When I say that the personality is not divine by nature, I mean that it is not infinite, not eternal, and that it has none of the divine attributes such as light and stability. That is the sense in which the individuality *is* divine. Nevertheless both are part of the same reality, God.

Let us glance at what the ancient Sacred Books said on the subject of good and evil. God declares in one of the Hindu Sacred Books, "I am both good and evil. All things were made by me." And in the Bible, "I am the Lord and there is none else, I form the light, I create darkness, I make peace and create evil. I the Lord, do all these things... I will not tolerate evil, I am indomitable, I will punish the wicked." Since nothing exists outside God, then evil (or what we think of as evil) is as much a part of Him as good. Are we to believe then that God is the Author of human sorrow, of war and devastation, wickedness and cruelty? Astonishing as it may seem, it is so.

We need considerable light to be able to accept this apparent contradiction... that God created evil and yet strives continually to overcome it!

Once I suggested that perhaps God created man as a sort of diversion, that perhaps He was bored and sought distraction, and now He sits back and with vast amusement, watches the way humans carry on! This does not alter the fact that He alone exists, that Creation is part of Him and nothing created exists outside of Him.

Now, let us look at the way our lower nature, the personality, was formed. Originally, it was a secretion, an emanation of the Spirit. When the Spirit wished to express itself it had to form an appropriate vehicle for each one of the spheres through which it had to pass in its descent to earth... vehicles which we call bodies. Going from the subtlest and finest vehicle down to the densest and coarsest, these are the Atmic, Buddhic and Causal bodies which constitute the higher Self, the individuality, and the mental, astral and physical bodies which constitute the personality or lower nature. The mental body (seat of our thoughts), the astral body (seat of our emotions) and the physical body (the material plane) are reflections, respectively, of the three higher bodies.

Someone will ask, "But why, if the personality reflects the individuality, is it so weak, why is it so blind and limited, so subject to error? The answer is that the individuality, which is common to all men, is divine indeed, all-powerful and entirely free in the heavenly regions where it bathes permanently in light, peace, and happiness... but it is unable to express itself in the dense regions of the material world except to the extent it is permitted to do so by the personality, or the three lower bodies. Thus it is possible for someone to be weak, ignorant and cruel on earth and at the same time to be full of wisdom, love and power on the higher planes... limited and ineffectual below, and quite the opposite above.

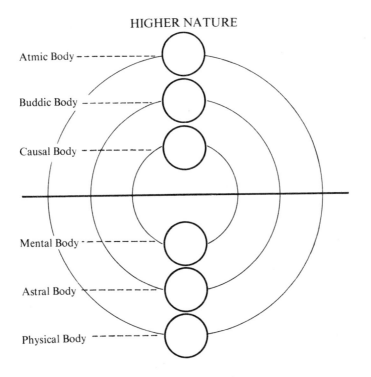

HIGHER NATURE

Atmic Body

Buddic Body

Causal Body

Mental Body

Astral Body

Physical Body

LOWER NATURE

We are told by esoteric science that man is a being who possesses a great variety of riches, that he is far more complex than he appears. This is where the esoteric science differs from the official science, which says, "Man is what you see, you can study him, learn all there is to know about him, by dividing him into so many organs, cells, chemical substances, all numbered and designated by name... that is the whole man... no more, no less." Esoteric science recognizes the existence of other bodies in man beside the physical.

The individuality, wishing to manifest itself in the dense and dark regions of the personality, is for the time being extremely limited. It will take time and more study and experience, more development, for the lower bodies or personality to be capable of expressing the qualities and virtues of the individuality. When it does, the mental body will have become subtle and penetrating enough to understand the Divine Wisdom; the astral body will be capable of feeling more noble, more selfless emotions and the physical body will have the power to act as it wishes, with nothing standing in its way.

The two natures are not really separate. The individuality keeps trying to bring its good influence to bear upon the personality but, since the personality is interested only in being free and independent, it listens only to itself and seldom obeys the higher impulses. Although it is vivified, fed and sustained by the individuality, it remains opposed to it, and will continue to be so until the individuality finally is able to control the personality completely. Once the personality gives in and is submissive and obedient, at one with the individuality, it will be a marriage, the only real marriage, the true love. In esoteric science this union is described as "joining both ends". The personality with three heads, like Cerberus, the dog guarding the entrance to Hell, is at one end, and the individuality, the Trinity of our divine nature is at the other end. The longed-for fusion, union or marriage, is bound to take place sooner or later in every case, but when it will be is difficult to know, for no two people are alike. In the meanwhile, the task of the disciple is to learn, as he experiences the trials and tribulations of life, to submit the personality to the individuality, becoming the willing servant and the instrument of the Divine Will within himself. That is the goal, the aim of all the instruction, methods and exercises offered by an Initiatic School.

Most people however, would rather bow to the capricious demands of the personality, rebellious, self-centred, anarchistic as it is, because they are convinced that is the attitude to have in order to evolve. People who are more intelligent and advanced, who have lived through many experiences in their numerous incarnations, prefer the way of self-control, self-mastery and the loftier qualities of the individuality. Thanks to their willingness to obey the higher Intelligence, or thanks to the Light within if you prefer, they receive divine help in overcoming the chaos and anarchy of their instinctive nature. The Divine Will manifests itself through them, expressing Itself in the celestial colours, forms, rays, fragrances, and music that surround them.

The question is, when we know what to do in order to become liberated and we know how to gain complete control over ourselves and progress in our evolution, why do so many of us give in to the personality and allow ourselves to be carried away by it... why? Because of the fact that even when we choose a higher level of consciousness, it may still be a product of the personality! We must reach the superconscious level, the level of the individuality. The individuality is characterized by an expanded consciousness, we reach it only when we understand that all creatures, all life is one, that all living creatures form a unity in the ocean of universal life in which all are immersed. Then our feelings change, we are filled with feelings of joy, wonderment, infinity. Human consciousness has its roots in the three bodies of the personality, it is therefore limited. Consciousness is the result of thoughts, feelings and actions: the more we think and feel and act, the more developed our consciousness becomes, but as long as we are tied to the personality our consciousness is limited to desire, we believe we are separate from others, from Nature, from the Whole.

The reason for prayer, meditation and all the other exercises taught in an Initiatic School is to establish the communication between man's lower nature and his higher one, the personality and the individuality, thereby stretching his consciousness and lifting it so that it may perceive the Truth.

Now let us suppose you are looking at a prism. With the eyes of the personality you will see no more than an object, a three-sided transparent crystal from which, when the light hits it, seven wonderful colours come streaming forth. This is wonderful indeed, but it is still on an ordinary level of consciousness. That is the way the majority observe things. But once you develop the higher consciousness of the individuality, you do not see the crystal as an object separate from yourself, you put yourself *inside* the prism and enter its essence from within, you feel and understand its nature because you are *within*. This makes a great difference in the way you think about the prism. It is the same for plants: if your consciousness allows you to enter a plant, you become one with the life flowing through it, one with the plant as if you were it... you *know* its properties, healing virtues and potential. And the same with animals, if you can penetrate an animal in your consciousness, you become one with it (without of course losing your human consciousness) and then whatever the animal feels, you feel.

The kind of education and upbringing which people receive today does nothing to prepare them for life, the real life. Their vision is limited to form, dimension, distance, weight and length... a limited scope at best! To enter a higher consciousness, they must now enlarge their sight and pay attention to the individuality, the higher Self, where neither time nor space exists, where you are one with all creatures, where past and present exist in your soul as the eternal present, and you can learn anything you want to know that has ever happened.

The reason for all the dire events and problems encountered by man during his life is that he allows the personality to dominate him, he lives in the personality. A small minority only make an effort to go higher, further, beyond their limits, to see with the eyes of the Spirit. For them the results are different, they have other sensations and feelings, other conceptions. As you see it is difficult to describe these things; although they are perfectly clear in my head, I cannot find the words to describe them to you, they belong to the fourth and fifth dimensions. I have the same difficulty explaining them to you as I would have in explaining third dimensional ideas to those who live in the second dimension. The fourth dimension is inexplicable.

When I say the personality is not divine by nature, it is a manner of speaking, for certainly God is at the origin of everything. Suppose for the moment that you are a gold-seeker. If you are lucky you may come across some ore one day, but you still have to know how to extract the gold. The ore and the gold originate from the same source, but still they are not the same thing. All matter has the same origin. You will be able to extract the gold from the matrix if you know how, but it is another matter to transform the matrix into gold... or to transform the gold back into base metal. These changes occur before our eyes in Nature. You can melt a piece of lead to make it brilliant and shining as silver, but not for long... very quickly a dull film forms and the bright silver-like metal vanishes. Again it will reappear momentarily if you scratch the lead, to disappear as quickly as it appears. Right before your eyes, the lead becomes transformed back into earth-matter.

The personality also comes from God, "What!" you say. "If God's nature is so utterly different from matter, how could He form anything as dense and dark as matter, as the personality?" My answer is the following example: have you ever watched a spider weaving its web? God proceeded in exactly the same way to make the world. We can learn from a tiny spider how God created the world! "Spiders?" you say. "Are they so clever?" I have no information as to how many diplomas are awarded annually to spiders, but if you will take the trouble to observe one as it spins its web, you will, if you understand what it is doing and for what reasons, be able to draw some wonderfully philosophic conclusions! The spider's web is a mathematical, geometric, perfect construction, a whole universe! How does the spider do it? First it secretes a liquid which hardens, forming a fine tenuous elastic thread with which the spider then builds its web.

Snails are also very instructive. One day I stopped one, "Tell me, dear snail... I am not stopping you for the usual reason, to cook you for lunch... no, I am seeking information. Tell me, why do you carry your house around with you on your back?" "It is the most economical way." "Isn't it tiring?" "Not at all, I'm used to it." "What made you form the habit?" "Oh, I don't trust people, I'm afraid that if I leave my house, a stranger will get inside, and that will be the end of me for I have no arms to fight with, I am too soft, too delicate and helpless. And so I avoid danger by carrying my house around with me." "Oh", I said. "This is a whole philosophy! But what is your house made of?" "Of saliva. The saliva hardens when exposed to air, and that is what I build my house with."

You see how interesting it is to converse with snails, besides which it helps you to understand how God created the world. He emanated a very fine subtle matter, His own essence, which then solidified. Between yawns you are thinking: silly bedtime story! One day everyone including the most learned and erudite will stay awake to hear such stories.

On the surface the snail and its shell are two separate things, but in fact they are one and the same, for the snail himself secretes the matter his house is made of. It is exactly the same, for the personality and the individuality, although the personality is dark and heavy and thick, hard as a shell, and the individuality is light and alive, quick and mobile... as different as night and day! Yet originally they were one. The individuality must form a vehicle, an envelope for itself exactly as the snail forms its shell, by emanating a substance which hardens... into the personality! We carry our physical body as the snail carries its shell, and we live inside it. The trouble is that man has never been told that he must not identify with this shell, his body, but rather with the power that formed it, the Spirit, the individuality. For that reason he remains weak and powerless, limited and mistaken. The body is not the man, the man is not his body! It is available for his use as a car, a horse, an instrument, a house. Man is pure Spirit, not matter, but all-powerful, infinite, omniscient Spirit. If he will identify with the Spirit, he can become outwardly what he already is inwardly, enlightened, powerful, immortal... divine!

You should understand that you are all divinities. Yes, divinities, and you live on a high plane free of limitations, shadows and darkness, sorrow and suffering, in the midst of abundance and joy. Do you want to know what prevents you from manifesting the splendour of those higher regions here below? The personality. Your personality is too unadaptable, too self-centred to capture the subtle messages from those regions... like a radio that cannot pick up all the stations. The waves and vibrations released by Cosmic Intelligence in the higher spheres are swift as lightning, and the matter of the personality is too dense, too hard of hearing to vibrate in tune with them, and so it cannot seize the divine messages. They flash by without making an impression and we continue to live in ignorance, far from knowing or experiencing the wonderful joy of our higher Being.

There are ways of changing this situation. If you choose to lead a pure life and become once again a child of God, then your heart will open and become generous, your mind will clear and your will become indomitable. The personality will become the willing instrument with which to express the divine life of the individuality more and more fully and correctly... until the day comes when both the personality and the individuality become fused with each other, the personality ceases to exist and becomes one with the individuality.

Until you can bring about that unity, you will go on receiving occasional revelations, but they are momentary, you will have intuitive glimpses of the dazzling reality, but they will not last, the clouds will return. Then again whilst reading a book or contemplating a landscape, whilst praying or meditating, you suddenly are aware of living a great moment. Again it does not last. A man's life is made up of such alternations between light and dark. When at last he becomes a true expression of God, a divine manifestation of Reality, then he embarks on the new life, he is *reborn.*

Someone will object, "What nonsense! There is neither rhyme nor reason to all that... I don't believe any of it." Well, let them go on leading the life of the personality! Sooner or later they will realize how much time they wasted. Far better to make a break immediately, yes, and believe, and start at once to practise self-control, to learn to dominate your instincts. Then you will go forward! But this does not mean you will be divine at once. No. You try and you fall, you try and fall and get up, you alternate between discouragement and confidence, until finally you allow the divine impersonal consciousness to take over and install itself inside you.

Sometimes we are weary and begin to doubt. There are so many bizarre philosophies in the world, ideas that are contrary to divine tradition, and sometimes we feel like forgetting it all and going back to the ordinary way of thinking. Those are times during which we must be very careful, very vigilant; we must realize what is waiting for us if we do slide back, and say to ourselves, "I feel this way because I am tired, that is why I have lost my appetite for praying and meditation and even reading... but it will pass, I will wait for this moment to pass." Everything passes. After the spring comes summer, and after summer comes autumn, followed by winter... and then spring again! Why wouldn't it be the same with us? Tell yourself to let the wintry moment go by, because spring will follow in no time. That is the way to reason, not to abandon hope as many do, and let go... which only makes matters worse, for it is very difficult to recapture a state of light and peace once you give up... very difficult indeed.

The fact is we have no choice, we must learn to cope with the personality and continue to work with it until we dominate it. Never forget that it must not have the last word *ever*. If we keep working diligently toward the High Ideal, things gradually change of themselves, we become recharged, our strength is renewed, the bad days are behind us, Spring is here, rivers are flowing, birds are singing, flowers fill the air with perfume, and all is well!

If you will do as I have been telling you, even when you are tired and listless, even if you lose hope you will still emanate a particle of light, a radiance, something sweet and gentle and harmonious. If not, you may be superficially strong and vigorous and sure, but, if you are one with the personality, then inside you are musty and mildewed, full of dust.

Videlinata, 23 February, 1966

Chapter 2

Jnani-yoga

The fact that man has both a personality and an individuality and spends his life indecisively alternating between the two, is clear and easily understood. It is when this situation affects our daily lives that we find it more complicated and harder to understand. There we need the help of knowledge, a science to tell us what is required of us, what rules to follow, how to classify people and things and how to discern what should be discarded... without this knowledge the difficulties tend to accumulate. It takes years of study and effort to subjugate the personality and make it work for you rather than against you.

It is a question of finding the right methods, effective and rapid ways of dealing with our conflicting natures. We may debate as to whether Heaven and Hell, Angels and devils, exist or do not exist, but one thing we can never doubt is the existence of two opposing natures in each one of us. The personality is not lacking in qualities, but as a counsellor, a guide, it is extremely dangerous, egocentric in its decisions, hard and cold in its judgments, interested primarily in its own concerns, vindictive, vulnerable, susceptible, demanding. It lacks both love and wisdom, and wants only to acquire and dominate. The trouble is that like a rich old lady it holds the keys to all the treasures, the cupboards, the larder! And we, impressed by its wealth, indulge its whims and give it its own way.

Yes, that is the plain truth. Man would rather be a slave to the whims of the personality than listen to his higher Self, languishing in a corner, neglected and despised. All of us, without exception, have a divine nature but, as it is stifled and trampled on, is it any wonder it remains silent? Every now and then the poor thing will try to get a word in, without insisting, without violence but, as most men prefer noise, the din and uproar of the personality drowns the gentle voice of the individuality. They do not realize what bad counsel they are receiving, nor that the personality invariably turns them against others.

I said that the personality is like a rich old lady who holds the keys to the cupboards, etc... because the personality has great wealth stored underground: the raw material of its instincts, appetites, passionate desires, strength and power but – and this is the trouble – it wants everything for itself. Capable, clever, resourceful, competent as it is (even with all its egocentricity) it nevertheless is completely lacking in honesty, morality, goodness, kindness, generosity, impartiality... that is why the personality is linked to the animal kingdom. Nevertheless it cannot be entirely bad, for it increases and protects all that man possesses.

The individuality is characterized by its many virtues; everything generous, good, noble and spiritual in man comes from the individuality. Not many know of its existence, examples in history or literature and art are not studied as examples of individuality, and living examples to follow are few and far between. We neither understand its language nor seek its counsel. To us it is vague and remote, and that is why this higher nature, so rich and magnificent, remains the prerogative of a small minority... whom the crowd looks upon as simple-minded misfits.

At one time in our history humans were not sufficiently evolved to grasp abstract ideas and the Initiates taught the divine philosophy in the form of allegory, the lower and the higher nature were represented as an angel and a devil. Should we still believe that we have a devil at our left and an Angel at our right? I believe they are there, but perhaps not quite in that form. We all have the two natures, whatever they may be called, the only difference is that some people give all their attention and every possibility to the divine nature (and are therefore better counseled and guided, comforted and upheld, enlightened and protected) and there are some who concentrate on the personality and obey its demands. Those who let themselves go to their instincts and appetites are inspired by the personality to do all kinds of things that incur drastic consequences. "Tempted by the devil," is the way they explain it, but the truth is that they have given in to lower tendencies over which they had no control.

The personality invariably creates complications. This is so true! A glance at human affairs shows that in all realms, social, emotional, political, the difficulties arise from the fact that our behaviour is the result of an egocentric, self-centred, personal point of view. Our standard and rule of life is to *take,* and it is this need to take that is the cause of all the insoluble problems man has to face. If people were better informed, better guided and counseled, if at least they accepted to be guided by conscientious and wise leaders if not by the higher Self, they would avoid some of their costly mistakes and no longer be miserable and discontented, in the dark, on the verge of suicide, or seeking vengefully to destroy the world... for the pleasure of seeing it burn.

The world would be a very different place if humans could bring themselves to listen to their higher Self. No one who is at all sincere can deny that his higher Self does its best to stop him when he is about to do something destructive... never with violence, never with loud trumpets, but gently, delicately, without encroaching in any way on your freedom, it whispers its advice. As we usually lack discernment and are surrounded by noise, we fail to notice that the higher Self is trying to tell us something.

The personality on the other hand, doesn't hesitate to impose itself violently, it will do anything to get its own way; at any time of the night or day it asserts its rights. Besides which, it knows very well how to send an impressive delegation to convince you with philosophic arguments that you, poor thing, are lost without its advice. What is more, it succeeds! People are often fooled by their inability to tell which of the two natures is counselling them. Some time ago I gave you the standards to use to be able to distinguish which of the two natures is speaking, but few of you bother to refer to them. Those who do benefit accordingly, and the others go on following the dictates of the personality, persuasive as it is, without even realizing it. The personality excels at intrigue, ruse, scheming and plotting, wangling its own way at all costs.

The great mistake men make is to identify with the personality. When they say, "I want (money, a car, a woman)... I am (ill, well)... I have (a desire, taste, opinion)," they think they are talking of themselves, but they are wrong: actually it is the personality which wants, thinks, suffers, and they obey it. As they do not analyse themselves, they don't know themselves, their human nature, and the different bodies of their being, the different planes on which they evolve. They identify without ceasing with the personality and the physical body. A disciple knows better, he knows that the real *"he"* is not the physical body, he knows that his instincts and desires are not the real self but something outside himself. This certainty allows him to progress.

In India the yoga of self-knowledge is called Jnani-yoga; those who want to find themselves and discover how to "Know Thyself", practice Jnani-yoga. The disciple learns who he really is, where he is, and that even if he were to lose an arm, a leg, his real self would still be intact, that therefore *"he"* is not his legs or his arms or his stomach, etc... *"he"* is something more. Glancing at his emotions he sees that there too, his feelings are not *"he"*, since he observes them and analyses them from a higher level. The same for his thoughts and opinions: "Are these *my* thoughts and feelings?" he asks himself. And again he finds that *"he"* is other than his thoughts. In this way he soon discovers that the Self he seeks is the Self that is above all, the higher Self, God Himself, all-powerful, luminous, omniscient. In time, after many years (even then it is not given to everyone), he joins his higher Self and becomes one with that Self. Now he sees how changeable, vulnerable, insignificant is the physical self, since he can do without it and discard it at will like a used envelope, whereas the real Self goes on existing.

Another aspect: a child says *"I"*, *"me"*. When he becomes an adult, no longer a child, it is still *"I"*, *"me"*: he has changed but the *"I"* and *"me"* remain constant. What is it that does not change with the body, with the emotions, with our ideas and thoughts? The Self! Our Self is always the same. Carrying our self-analysis a little further, we see that our Self is part of God Himself, and from then on we make every effort to bring our Self to Him consciously. We see that our personality is not eternal but a mirage, an illusion, a mere reflection of the real Self which the Hindus call Maya. The danger of Maya, or illusion, is that it lures us away from the divine Source, our real Self, into the idea that we are separate entities, myriad independent beings each with his own tendencies, desires and emotions. But we are not separate from each other, nor is it the world that is Maya, as some people think, but the personality, the lower self which is always trying to convince us of our separateness, not only from others but from the universe, the Creation and all its creatures. The world is not Maya. Being matter, the world is real, as lies and deception and Hell are real. Maya is the illusion that we can

exist separately, separate from universal life and from the one
Being whom we neither feel nor perceive although He is
everywhere, because the personality prevents it. When,
through study and meditation and, above all, through identification, we begin to rediscover who we are, who our real Self
is, then we will also begin to understand that it is not a question of multitudes of beings, but rather of one Being working
through all beings, animating them and manifesting Himself
through them (whether they are aware of Him or not), a single
Being controlling and directing all our manifestations. When
we understand this greatest of truths, we will no longer be
able even to consider declaring war on each other, everyone
in the world will be part of the one collective Being.

An illustration for you : suppose I place several glasses on
a table, all different as to their substance, form and colour.
Now, I pour perfume into each glass, the same perfume in
all... the containers will be different but the *contents* will be
the same. The glasses retain their static, unchanging form
while the perfume, the essence, rises and spreads fragrance in
the air above. Perfume is etheric and subtle, it mingles and
blends with the air : a fusion takes place above and the contents of all the glasses are as one.

This image illustrates the fact that mankind is not divided
into separate beings as the personality would have us believe.
If you accept the illusion, if you are resigned to living with illusion, you will always be mistaken concerning Reality, it will
blind you, deceive you, keep you from the Truth. Or rather,
let us say that the other philosophies are true as far as matter
and form (the container) are concerned, but false where the
contents, the soul and Spirit – or unity – are concerned.

Now let us say that several people come and sit down at
the table who all have one thing in common : they love and
respect each other. Physically, superficially, they appear to be
separate, but that is the limited point of view, for the currents,
forces and energies that circulate between them, the fusioning

of their ideas and emotions, the love they exchange, raise them up *as one* onto a higher plane where indeed they are one. Of course if we confine our sight to the glasses, the physical bodies, the containers, then we see only their form or outline, their dimensions and colour, but the outlines and barriers vanish if we concentrate on the contents. It cannot be said: Here is where life (the perfume) begins and here is where it ends. You cannot possibly limit a live and moving substance that keeps changing subtly, that keeps radiating.

Someone wants to make a drawing of me, for instance. Is this body, this face or profile the real *"me"*? Am "I" my physical body, does the "I" have any limiting contours? No, "I" am not what you see, "I" am not the physical body, "I" am something else that thinks and feels and acts... which may possibly be a bit more substantial than what you see before you!

And take the sun up in the sky with its determined form, shape, dimensions. How is it able to affect us, to reach out and touch us from so far away? It touches us by expanding, dilating itself, stretching out to reach us through space. If the sun can do that, could we not do the same? Yes, by sending out the rays of our thoughts. By thought you can reach anyone you wish, no matter how many thousands of kilometres away. What are our thoughts made of? Our emanations are our quintessence, as the sun's rays are the sun's quintessence. The sun's rays reach not only the earth and its inhabitants but further, many thousands of light-years away, all over the universe. The sun sends out into infinite space its very soul in the form of sunrays; they are its thoughts, and thoughts *are made of the self.*

Take the planets: on the earth for instance, there is more liquid on earth than solid ground, more gas or atmosphere than liquid, and the etheric portion of earth can reach further than the sun. This is also true of Mercury, Jupiter and Venus... which shows that all the planets that are adjacent are actually touching, intermingled, forming a single unity! Outwardly separate and far apart, inwardly (in their subtle side) they are not only adjacent but actually overlapping. In the same way we too are connected, overlapping through our thoughts and emanations. The Bible tells us, "Ye are gods". Men are gods! If all too often they resemble animals still devouring each other more than gods, it is because they are too steeped in the personality, mired in the regions where they are limited, cut off from others.

The truth is, we are all one. That is reality. If you think about that before deliberately hurting someone, you will see that you hurt yourself by hurting him, he lives in you and you in him. True morality lies in understanding that when we harm others we are in fact harming ourselves. We already know that when the one we love suffers, we suffer too, as if we ourselves received the blows, and when something wonderful happens to someone we love, we rejoice as much as he does, his triumph and joy is ours as well. This is the result of the philosophy of unity, it means we understand the philosophy of love and universality. Until we do, we are more apt to rejoice at our neighbour's misfortune... alas yes! We secretly rejoice at the difficulties the *others* have to face.

That is what the personality does, maintains us on a lower level of consciousness. We do not realize it because we identify with the lower nature and not with the higher Self. To convince our instincts and desires to change is the most difficult thing in the world, but we must at least be conscious of where they come from, that it is not our true, higher Self dictating to us. If you become detached from the personality you weaken it, and it is then easier to establish the bond with the higher Self... and identify with it.

If you study the personality, learn its ways and methods and weigh its advice, you will see that it cannot hide; its demanding ways, its aggressive way of hitting anyone that stands in its way, of yelling threats and abuse, are all too evident. Once you recognize its tricks, you can never be fooled again, but you still have to be vigilant and careful. For instance, you may decide to give up something it likes... but it is waiting for you at the next turn, presenting the situation in a better light, convincing you that you are wrong to give it up! If you enjoy tobacco, wine, women, money, the personality knows exactly what to say to justify your likes... the very day you renounce something, there is the personality smiling and understanding, "So, old chap, you have resolved not to drink anymore? Wonderful! This calls for a celebration!" And off you go to the corner pub to toast your new resolution! Is the personality not extraordinary? Such guile, such ruse, such wiliness! Yet it must not be done away with completely, it is part of us and we must learn to live with it, primarily because it is so useful, so vitally necessary to us as the reservoir of all our potentialities and assets. We must simply be more intelligent than it, we must know how to make it submit and obey, to put it to work, for there is nothing and no one as capable as far as work is concerned, no one as tireless, devoted, capable. But if you do not know how to make it obey, if you cannot control it, you will be the one to be controlled and there will be nothing left of you, not one crumb!

To come back for a moment to the glasses and their contents, the perfume or essence, I cannot understand how humans have gone on for such a long time without seeing the truth, which is that we all have a soul, a quintessence or the "contents". How could we be so stupid as to ignore the best in us, all that is alive and intense, and devote ourselves to dead matter? If you concentrate on matter, you come to identify with it, nothing is more dangerous, for you too become inert, frozen, unable to escape from your enemies. You must learn to move, to change places, to fly away like the birds into the sky where you are no longer at the mercy of circumstances. No one can capture you if you keep moving, if you are alive, subtle, weightless, transparent and inaccessible, you cannot be seized, you soar above in safety. "What about the physical body?" you ask. The physical body is without a doubt heavy matter, and therefore exposed to danger, but the soul is not: just try to seize a man's mind or Spirit, his conscience. Man's soul is above conditions and circumstances. The *glass* can be seized and held but not the perfume floating in air!

With what ease we repeat the activities of our daily lives, concerned as they are with concrete things such as food, drink, all the objects available to us... but to repeat and re-experience a moment of ecstasy and rapture, is much more difficult. Ecstasy belongs to the invisible world, outside the reach of humans. Everything in the world of the personality can be repeated endlessly, the acts, quarrels, ceremonies, comedies and tragedies... but in the higher world of the individuality it is not the same.

Anyone who sinks too far down into matter is from then on at the mercy of other entities who dispose of him as they see fit, he is sent here or there, apprehended and... killed, if that is what they have decided. I have drawn this conclusion: if you want to be above all circumstances, safe from danger and all attempts on your life or liberty, safe from misfortune of all kinds, what you must do is fly above all the world's events, be always climbing higher and higher, and never let yourself become crystallized. In that way no trouble, no loss or failure, no wars or revolutions can touch you because you are above, inaccessible, out of reach, far, far above.

That is what the Initiates understood. How many times in history were they thrown in prison, threatened with death to make them talk... not once did they betray their secrets. There was one who was told his tongue would be cut out of his head if he refused to answer, and so, rather than betray what he knew, he bit through his own tongue with his teeth, and spat it out at the feet of his tormenter as proof that nothing could change him. Yes, history is full of such examples. We would do well to think about them.

Le Bonfin, 26 July, 1968

Chapter 3

Giving and Taking

I

Today I will go on giving you illustrations to show you the difference between the personality and the individuality... it is a world in itself. The best image comes from the word *persona*, the Latin word for the mask worn by actors in the theatre, a different mask for each role. The actor's personality changed with his mask and costume, but he himself remained the same throughout the entire play.

The theatre gives us the explanation for what the personality is: the role each man plays during each life. A human being who is about to incarnate dons his mask and costume, his personality, for the role he must perform, a role which changes from one life to another. His mask makes him appear as a man or a woman with certain virtues and vices, with such and such opportunities and drawbacks. In his next incarnation he will have a new role to play, another personality, but he himself, the real Self, will remain the same. The personality alone changes, it is passing and temporary; the individuality is the result of all his lives, his apprenticeships, it is his heritage, the fortune he has acquired during his life. His wisdom and his qualities and attributes are his throughout all his incarnations; his personality changes with each incarnation.

The individuality can be compared to the sun, the personality to the moon. The moon passes through different phases, changing ceaselessly; it is never still, nor has it a light of its own, not being the centre of the planetary system as is the sun. The individuality is always radiant, always luminous and powerful, like the sun.

Yesterday I said, "The most important thing is to know who it is you work for." If you work for the personality alone, which is something that is always changing, never constant, which leaves no trace behind it, you will lose all your wealth, all the things you have acquired, they disappear into the void, because that is what you have been working for: the void, the wild wind.

Now let us see what happens as a result of working for the personality, and then look at what happens when we work for the individuality. The personality's most striking characteristic is the fact that it always wants to take, to grasp and possess. I have told you that the personality is a trinity, a triad in reverse, the lower manifestation of man's mind, heart and will. When the personality meets opposition, it rallies all its willpower, all its intellectual and emotional resources to the attack, in order to get what it wants. That is the way to distinguish the personality from the individuality: the former is invariably selfish, its tendency is to take, and to keep for itself what it gains.

The individuality does the opposite. It wants to give, to radiate, to emanate, to enlighten and help and sustain others, to project something of itself with great generosity and self-abnegation. It desires nothing for itself and is never irritated at having to share, rather it is delighted to be able to do something for others, to feed them, quench their thirst, enlighten them. With the higher triad of the mind, the heart and the will, through our mind the individuality strives to shine, through our heart it strives to encourage and warm all people, and through our will it strives to enliven and liberate all creatures.

INDIVIDUALITY

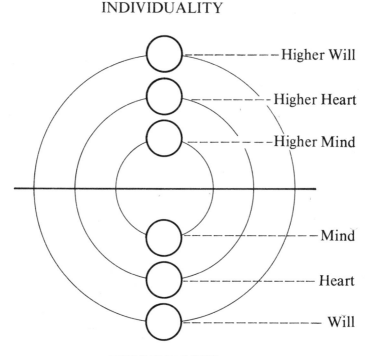

Higher Will

Higher Heart

Higher Mind

Mind

Heart

Will

PERSONALITY

A man has two different natures within himself and he is constantly being drawn toward one or the other, propelled sometimes by the geocentric force and at other times by the heliocentric force, or, if you prefer, the centripetal force and the centrifugal force. If he lets himself be influenced by the coarse, vulgar, shabby and trivial personality, he is carried off by the geocentric current. Unfortunately, that is what most men do, unaware that this current will lead them to Hell.

Everyone finds it perfectly natural to work for himself only, without a thought for others, but rather seeking to crush and dominate them... or get rid of them entirely. This attitude is so prevalent that if you try to tell them they should behave in any other way, people are astounded and wonder which planet you come from. It is considered quite normal to lie and cheat, to take advantage of others, to connive against people, to use them with no consideration for their own well-being... what is wrong with that? But the consequences of this attitude are not great, for the man who follows his personality is forced to disobey nature's laws and in so doing becomes unpleasant, arrogant and cruel. And of course, if he goes too far, if he goes beyond the limits, there will be repercussions, counter-actions against him, he will have to go through a good many hard lessons until such time as he realizes he must not serve the personality exclusively. Yes, the consequences are not pleasant, sooner or later such a person will be in trouble, for his family, friends and associates will despise him.

On the other hand those who are aware of the existence of the higher Self try their best, they make every effort to enter the heliocentric current. The foremost quality of the higher Self is to give, to radiate. The truth is that all virtues are no more than a radiance, a projection from the centre toward the periphery, a need to take something of oneself and give it, to sacrifice oneself. A man who manifests such a need is obliged to conquer certain weaknesses of the personality such as cowardice, fear of pain and suffering, and death. Fear, or lack of courage, is conquered by the desire to give, to radiate, to flow as a spring flows, with love.

In order to understand the personality one must study the moon and the earth: the earth seeks only to take, to absorb and, unlike the sun, it contributes nothing to the Cosmos. If the inhabitants of Jupiter or Saturn ever train their telescopes on the earth they may see a feeble light, like the light of the moon or the other planets, but even that light is not its own. The earth is not capable of producing light because of its egoism, it is still too selfish and self-centred: egoism does not project light. Light comes as a result of something you tear out of yourself from your innermost recesses, a sacrifice, a manifestation of love, of self-abnegation.

The sun is a living illustration of the tendency to give, the desire to give, and the earth illustrates the tendency to take. This does not mean that the earth gives nothing at all, it produces flowers with what it receives, and fruit, but for whom? For the earth, for itself. Do you think that other stars benefit from the earth's flowers and fruit? No, only the earth itself gains and possibly its children, which is the same thing. The earth takes and then goes to work with what it has taken, but the earth keeps its products for itself. The personality does the same, it takes and keeps everything for itself. The sun keeps nothing, it sends itself out into space so that all creatures may benefit. Those are the two laws: the law of taking or absorption, and the law of giving, of radiance. The sun is the perfect model of the law of radiance, of streaming forth, of emanating.

We watch the sun as it rises; what do we see? A perfect example of the individuality, of the Spirit in the act of giving with the divine generosity of the Deity, Almighty God. We watch and watch... but as no one has ever explained or interpreted for us what it is we are watching, we go on all our lives without learning more than we know already... the earth's laws, the laws of the personality. In other words, we go on *taking.* Once you know what the sun is, you begin to feel its power and grandeur, its majestic, tremendous *giving,* and you cannot but work to change everything in your way of thinking. If you understand this, you will soon be rejoicing because little by little you will start to resemble the sun!

That is what everyone must now learn: to give, to give without waiting for any recompense or recognition, as the sun gives. People make little sacrifices and then look for thanks or praise or a compliment... it is considered normal to do so. But those are the earth's laws, not the sun's.

The path I am suggesting is not for anyone but disciples, the children of God, who want more than anything to grow to be like their Heavenly Father. I am speaking for them. To resemble the Lord, there is no other way than to strip oneself completely, to snatch something from one's innermost heart... to make the supreme sacrifice, of giving one's life. Of course, many people give a penny here, a crust of bread there, old clothes, shoes with holes in the soles, thinking this kind of charity will take them straight to Heaven. Don't you believe it! If it were so easy... no, until we have learned to sacrifice some of our weaknesses, some of the things we covet and never deny ourselves, we do not know what sacrifice is.

Egoism has a most pernicious effect on us. When we refuse to give and keep everything for ourselves, it has the effect of stopping up our inner channels. You know what happens when a stream dries up, the fermentation produces a bad smell and attracts vermin. Yes, simply because the spring stopped flowing. And the same thing happens in human beings, the personality becomes like stagnant water. But on the other hand, there is nothing more resourceful than the personality when acting for its own purposes. Yes, when that is the case it is very active, very prompt, violent even. The individuality is not as dynamic or as resourceful, but it has an extraordinary way of always flowing with light, with life, with inspiration... it vivifies. The individuality is a spring, a source, and when it starts to manifest itself, spouting forth with an abundance of love, then kindness and purity and generosity and *light* sweep through you, and you feel cleaner and lighter, washed, radiant, pure!

You see, it is easy to evolve. "What," you say. "Easy to evolve? For years I have been trying, and I am not evolving." Simply because you are not working on the essential, you are not applying the law of sacrifice, of giving. What you do is always for yourself, to enrich yourself. Even when you are reading, even when you study, it is to take something. When you begin to distribute what you have learned in your books and in your life, you will change. People work of course, but always to take, to acquire power, to extend their influence far and wide, like a large company that opens branch offices to do more business and make more money, but not to *give.*

Jesus touched on this question, not perhaps in the same words, but if you know how to interpret the Gospels, you see that he was talking about the importance of detachment, of stripping oneself of all that is not important. When the rich young man asked him: "What shall I do that I may inherit eternal life?" "Obey the commandments," answered Jesus. "I have observed them from my youth," said the young man, "What more should I do?" "Sell whatever thou hast... and follow me," was the answer. When the young man heard these words he was desolate for he was not willing to give up his possessions, and he did not follow Jesus. If Jesus asked such a thing, it was because he knew the vital importance of the two laws: taking and giving. Why give? In order to be free, free to follow him and become a sun! You see, it is the same idea only in a different form.

If you understand what I have been telling you, you will from now on look at the sun with new eyes and great changes will take place in you. Everything depends upon your comprehension, the way you see things. A man who has a deep and correct understanding can release heavenly currents within himself and be transformed into a sun, giving ceaselessly and yet becoming richer all the time.

Jesus said, "If any man will sue thee at law, and take away thy coat, let him have thy cloke also". Why? To be like the sun! Jesus didn't actually say "to be like the sun", but it is the same idea: to become inwardly so strong and powerful as to be above all forms of fear. Fear must be conquered, for it says in the Gospels that the Kingdom of Heaven is not for the fearful. The personality is fearful, not the individuality. The personality is afraid because it feels isolated, vulnerable, miserable, it wants to take in order to compensate for its lack of security. Fear keeps us from manifesting love, love being incompatible with fear, but where love is, fear vanishes. The personality stops at nothing, it is capable of the most ignominious, disgraceful deeds... and that is why it must be shunned, avoided at all costs... that is all there is to it. We should not try to kill it but withdraw from any contact and give it the minimum of attention and sustenance, in exchange for a closer contact with the individuality, the higher Self, the sun.

Initiates also have to drag the personality around with them, the difference being that they do not pay attention to its demands and give it only enough to keep it alive. They support it but it is not the mistress of their house... like the sign in some homes: "I am the master here, but my wife is in command." The Initiate's sign reads: "The personality is my servant, I am its master." The Initiate does not kill the personality, nor starve it to death the way hermits and ascetics of the past did. They were taught to live in great privation, to beat themselves and wear hair-shirts, but that kind of treatment made the poor personality useless.

The personality should be cared for, fed and washed properly, but we should not allow ourselves to be taken in by its crafty scheming. Do you deprive a servant of food and lodging? If you have servants, you make sure that they have enough to eat and drink, but you do not let them run your affairs or give you orders. I know there have been cases where the servant became indispensable... she cooked such wonderful fritters for instance, that her boss, being a little too greedy, was unable to do without her... and married her for her fritters! Read the biographies of some of the world's famous men, and you will see how true this is. But I am speaking in general. The personality should not be killed but rather should be tamed, domesticated, supervised and not given too much freedom, otherwise during your absence it will invite all the neighbours in to drink your wine and eat your food... and when you come back the cupboards will be bare! Nothing but broken bottles and empty shelves! That is what the personality is perfectly capable of doing. When the mind is not in control, the personality joins its friends from the astral plane, that is, your lower thoughts and feelings, and off they go on a spree!

To take is the old way, the ancient teaching, and to give is the new way, the new Teaching. Of course when I say "give" it is not a question of giving to anyone and everyone, but to give to the luminous spirits above, the angels and Archangels, the saints and prophets, to God. That is where we should send our strength and force, our very lives. But how can a man give his life in that way when he is held back in the lower regions by the personality? It will do all it can to dissuade him, saying that it's an idiotic way to behave, and he will go on giving in to the attraction, the pull of the earth instead of entering the magnetic field of the sun.

Now, I will show you how you can apply Newton's law on the spiritual plane. Newton condensed the law of universal attraction into a formula: the planets move toward the sun in direct ratio to their weight and in inverse ratio to the square of their distance to the sun. The attraction is proportional to the weight of the bodies and inversely proportional to their distance. Later on scientists made the experiment of weighing the same object at both the pole and the Equator, and found that at the Equator the object weighed less than at the pole. Why? Because the earth is slightly flatter at the poles, the distance from the centre of the earth being less there than at the Equator. The attraction is hence greater at the poles and the object is heavier there. But the further the object moves from the earth, the less weight it will have, until it moves out of the terrestrial field of attraction and eventually becomes weightless. Let us suppose it then passes into the solar field of attraction, the same law will come into play to attract it to the sun : the object that was attracted to the earth is now attracted and absorbed by the sun.

The same thing happens to human beings, situated somewhere between Heaven and Hell, the sun and the earth, that is, between the personality and the individuality. If a man remains too close to the earth, his personality retains him and he weighs a great deal. But if he can draw away, the personality will lose its power over him until the time comes when it has no more power at all, and he is free to go directly to the centre of the sun, to his individuality. It is the same law, as you see, but the world's astronomers are too busy with the physical plane to investigate whether the same laws exist on the spiritual plane.

In order to be able to detach himself and voyage in space, man must cut the strings that attach him to earth and, to cut these strings, he must learn to give. That is what it means to draw away, to become detached. It means to sacrifice, to cast off our self-indulgent habits, renounce our desires, and develop generosity. By giving, you draw away from the earth until you reach the point where you are absorbed by the sun. For me, this is so simple, so clear. The difficult thing is to find the words to express what I see and feel inside me. When I have to search and search for the words and sentences to explain something so simple and easy, I no longer see it clearly! But I am talking relatively, I always see clearly... but not as clearly when I have to put things into words!

If now, after all these explanations, you still have no desire to embrace the philosophy of the individuality, it proves that the personality has a good hold on you. I know of course that for millenniums the personality has been influenced by the family, by all kinds of philosophies, and it has become so strong and resistant that, no matter what you say to the majority of people, they will go on listening to it and obeying its demands. They say, "We are all right as we are, and even if there are trials and tribulations to go through, what of it, that's life!" They accept their suffering and their slavery because they have nothing better to turn to.

— I know also that this philosophy will not be accepted by those who are so firmly in the grip of the personality that they have no taste, no desire for a higher life, a more beautiful, more poetic life. You say, "Well, if you know ahead of time that humans will not listen to you or follow your ideas, why do you go on speaking?" Because I know that some of them are not so firmly caught, not yet prisoners of the personality, it is for them that I speak; there is hope that one day those people will climb out of the circle of the personality and come nearer to the divine world. I have no illusions about the others. Later, perhaps, much later, after several more incarnations and much misery, many catastrophes and hard lessons, they too will be able to extract themselves from the claws of the personality.

The personality came into being at the beginning of the animal reign. Animals have a higher nature too but it is dormant. In human beings the higher nature is beginning to manifest itself, a little, whilst in the great Masters it is free to manifest all the time, it is the dominating factor. That is the ideal we should all have. How long it will take... never mind time. The sun is the symbol of the ideal, and for that reason we must adopt it as our model, our standard. Someone will object, "But don't you know the sun is not human, not alive?" Perhaps, but the sun *does* more than humans do. It is surely better to want to emulate something which may be unrecognizable by human standards, but which is nevertheless infinitely above anything man does, rather than to remain on the level of weak, selfish and ignorant people who prefer the dark and who may even be criminals.

Suppose the sun were merely made of rock, or metal in fusion... what difference does it make? None at all to me, for as long as it manifests the sublime, divine qualities that are way above human qualities, and brings us light, life, and warmth, then I will go on trying to approach it and to imitate it without wondering what it is made of... vegetable, mineral or anything else. If its qualities surpass human qualities, I will follow it because I know it will make me more intelligent, it will heal me and uplift me. Whereas being with humans is more apt to make one ill, discouraged and unhappy. Someone is bound to say, "Good Lord! this fellow is so warped, now he is saying the sun is alive and intelligent!" Well, why not? I am not the only one to feel that way, I am merely following in the footsteps of my renowned predecessors.

If you retain one thing only from today's talk let it be that you should try to give in a little less to the personality. And the best way to do that is steadily to diminish its rations. The Parable of the Unjust Steward shows us a steward enslaved by the personality... until the time comes when he realizes that another, better Master exists, and then he prefers to devote all his efforts, his time, to Him. That may not be the way this Parable is presented in the Gospels, but the meaning is the same. The Steward (or disciple) always worked for his personality until he began to reflect on spiritual things, whereupon he said to himself, "If I do not make friends with the Heavenly Beings (the individuality) on the other side, who will be there to receive me when I get there? I see now that I shouldn't have been devoting all my time and attention to this boss, I should give him only fifty percent, not one hundred percent of my time and effort, and devote the rest to the higher Beings." The Bible goes on to say, "The Lord commended the Unjust Steward because he had done wisely". Therefore the disciple, instead of giving his physical body everything it requires, should diminish its rations rather than increase them! All the energy he has been giving to his physical body to gratify the cravings of his lower self, should be transferred to the higher Self, the individuality.

We have been serving the personality far too long, it is time we were unfaithful! Instead of contenting it all the time, we should make it do with half or a quarter of what it is used to, and give everything it has had, all the time, energy, thought, feeling, pleasure-seeking efforts it has been requiring from us, to the individuality. And then when the time comes to leave this earth, preparations will already have been made, there will be good friends waiting for us in Heaven. I am not inventing these things, rather I am interpreting the words of Jesus... if you do not believe what I say, go and ask him and you will see. He will say, "These things are true, and your Master is free to interpret my words as he will. He tells the truth, he is never in error, and he has a right to put these Truths in whatever way he wishes. I agree with him... besides, this is the way he has always worked."

Le Bonfin, 28 July, 1968

Giving and Taking

II

Well, dear brothers and sisters, were you able to digest, assimilate yesterday's lecture on the personality and the individuality? This question of our two natures, the higher and the lower, must now become absolutely clear to you, because it is so important. Depending upon which nature we identify with we are attracted to Heaven or we are attracted to Hell: the personality is the link, the open door into Hell, the same door through which its inhabitants are able to get at us.

In another talk I told you about the two currents that circulate, one toward the sun's centre, and one toward the centre of the earth. Innumerable human souls are caught up in the subterranean current that acts through suction, dragging them toward the centre of the earth... call it the force of attraction, or seduction, or anything you like. Those who let this happen end up as demons, fiends, devils, whilst those who are carried along by the other current are swept upward, to the sun's centre. It all depends on which you choose, the personality or the individuality.

There are many thousands of currents circulating in Nature, but we will discuss only two, the current of light and the current of darkness. Initiates, philosophers, visionaries of all kinds have presented the same idea in many different ways, trying always to prove that the world is divided between two main conflicting forces. This is what the Persians believed, with the god Ormuzd representing the light and the god Ahriman the dark, the shadows. But this distinction has not always been properly interpreted and Manicheism has been greatly misunderstood. The Persian Initiates were reluctant to indicate evil as the principle that opposed good, without the good being able to conquer evil. They simply preached the idea of polarity, with God as One, but when He manifests He becomes Two, masculine and feminine, positive and negative, luminous and sombre, and with darkness not considered necessarily evil.

Nature shows us that the dark contains forces which work for growth... night, for instance, is night evil? The unborn child begins his life in the dark; for the first nine months he is in darkness... is that evil? Light and shadow are symbols of good and evil, like the right and the left: often the right side is associated with the good and the left side with evil, but actually it is only a manner of speaking. Man has two sides, but he is nevertheless one single indivisible entity. Look, if your left hand slaps your right hand, it doesn't mean they are two separate entities at war with each other. No, it is the same person who does both the hitting and the receiving. Once you understand the law of polarity you will have the answer to a lot of things you now consider mysteries.

In Nature, everything is good. I am not saying that devils are good, no, but one day, when they have become tame, appetizing and well-seasoned, they will be a feast for us! It says in the Talmud that at the end of the world the monster Leviathan will be cut up and salted and served as a feast *for the just*. What a privilege if indeed we are among the guests, to dine together on the Monster at the same table! If the Leviathan, who is certainly no better than devils, is to be served as a feast, why not other devils? We may eat their flesh one day... but at the moment it is they who eat us.

The personality links man with Hell, and the individuality links him with God. Which is why, if he lets himself be swept along by the luminous current that puts him in harmony with the individuality he will rediscover his true divine essence. What I am telling you here fits in with all esoteric and religious traditions, which always urge man to go back to his origins, to find himself as he was in the beginning... at the moment he is somewhere between Heaven and Hell, having erred for so long and changed his condition and habitat so many times that he no longer knows where to go. He has forgotten the Ancient Wisdom he once had, and that is why he now needs to be guided. In the remote past he was guided by his own light but now that he has lost it, he no longer knows which direction to take. One does meet people occasionally who remember what their origins were, they know where they come from and where they are going, always guided by their own inner light. Each day proves the veracity, the truth of what they receive, and their conviction and certainty grow accordingly, for light is never misleading.

But except for a small minority of wise men, most people live in the midst of uncertainty and anguish, wondering what the meaning of life is, what will happen to them after death. The consequence of this kind of thinking is all too obvious in what we read in literature, what we see in Art. How often is our last shred of faith destroyed by a film, a piece of art work, a book? Unfortunately the younger generation prefers such books to anything written by wiser, more luminous beings, divulging their discoveries, their inner experiences. Writers, philosophers and scientists have done their best to annihilate the few rays of light left to mankind, and now no one knows what direction to go in. That is why once again we need to be shown the path. Those who understand will choose that path, and others, once they have experimented and tasted everything, will also understand... but it will be much later.

I was telling you yesterday that the most striking tendency of the personality is to take, and this even at the expense of others. The personality has no morals; no pity nor compassion, nothing. It feels entitled to everything, it absorbs all it wants and never seems to have enough, it is never grateful, never satisfied, quite the opposite, no matter how much you give it, it is vexed that it wasn't more. Along with this need to take, always to take, the tendency develops to be cruel, vengeful, jealous, rebellious.

The personality thinking only of *taking* can be likened to the earth, as I have told you. The individuality which does nothing but give is like the sun, and the human being is somewhere between the two, sometimes behaving according to the personality, sometimes the individuality. I also alluded to the moon, but so rapidly that I would like to go over it again and add more to what I said. What is the moon? A globe, an earth like our earth. Speaking scientifically, astronomically, it is not really like the earth, but symbolically it has the same nature as the earth, for it also takes. The moon and the earth are feminine in their function, they both take but each in its own way, and you will see how, from the Initiatic point of view. The sun gives; the earth and the moon take. Never mind the other planets for the moment, we will simply take these three symbols, the sun, the moon (represented in the Ancient Mysteries as the masculine and feminine principles) and the earth. The earth represents the physical body, the moon represents the astral body or lower soul, and the sun represents the Spirit. Mercury represents the mind or intellect, but we will not discuss that today.

Now let us think about the sun, the moon and the earth. The sun, that is the Spirit, gives and never takes, whereas the soul takes, and the body also takes. In what way do they take? That is the interesting thing. We are not to think of taking as necessarily evil, for there are ways of taking. You can be entirely selfish in your taking, brutal, greedy; or you can take calmly, as still water takes, as a mirror, the soul at peace, reflects Heaven.

And so in this way you can classify human beings: those who are under the influence of the earth and take egotistically, without love, without pity, and those who come under the influence of the moon such as mediums, visionaries, poets, mystics, who try to take, or rather, capture certain things from other regions, or from the sun, in order to reflect them. To reflect something is a way of giving. The soul of the medium is so receptive it receives revelations from above, and the medium then transmits them to others, makes predictions and so on. Someone who is receptive and sensitive enough to capture the waves and messages from above is not the same thing as someone who swallows everything like a bottomless pit. A mirror gives nothing of itself of course, but at least it reflects, it sends something back. And that is the moon's function: to reflect. As it is situated between the sun and the earth, the part that is turned towards the earth reflects what is lower, and the part turned towards the sun reflects what is higher. The moon waxes and wanes as if to show that during a certain period it reflects everything evil and infernal, and during another period everything that is heavenly and good!

The human soul is closely linked with the moon, the physical body is linked with the earth, the Spirit is linked with the sun. And like the sun, the Spirit never takes, it only gives and gives, gushing forth inexhaustibly. Because it is always emissive it cannot receive. The earth cannot give, it remains content to receive all the time and with what it receives it produces flowers and fruit, but the moon is at times under the influence of the sun and at other times under the influence of the earth. That is why people who are influenced by the moon are so unreliable and unstable; they are full of poesy, they have inspirations, and all of a sudden they fall into a deep depression! Then again they are full of enthusiasm. You never know what is going on with them, and one should be careful, their intuition is absolutely correct at times, and then again it can be totally misleading. The moon is the region of lies, illusion and deception, but it is also the realm of clarity and purity... why?

In studying the Sephirotic Tree you see that starting with Malkuth (the earth) and moving upward in a vertical line, you come to Yesod, the region of the Moon, and then Tipheret, the region of the Sun. The moon being situated between the two is influenced by the Sun in its higher region or half, and in its lower half by the earth. Everything that comes from the earth, that is, lies, clouds, terrifying visions, images, phantoms are reflected there. But once above this deceptive zone of the moon you come to the side that is influenced by the sun and filled with light. That is where all true clairvoyants can be found, in the higher region of the moon.

There are two categories of people in the moon, those who are emotional, unstable, bizarre and a little lunatic, and those who, on the contrary, are lucid, clairvoyant and pure. For the moon can also make you pure; if you want to be pure, to become pure as limpid water, link yourself with the higher side of the moon, for the moon rules over both crystal-clear water and polluted water; all water, dirty or clean, is influenced by the moon. Hecate represents the lower aspect and Diana, the chaste Diana, represents the higher aspect... and also Isis.

Now as to the two words, emissive and receptive, which I use frequently, I will add a word of explanation. People who are only receptive absorb everything good or bad without discernment, and thus receive the impurities and illnesses of others. Receptive people, such as mediums, are extremely vulnerable because they do not know how to defend themselves, and are at the mercy of every influence from all entities. Among people who are emissive, or solar, are the mages. With mages it is the masculine principle which dominates, that is, willpower, the desire to give, to construct, to build, to

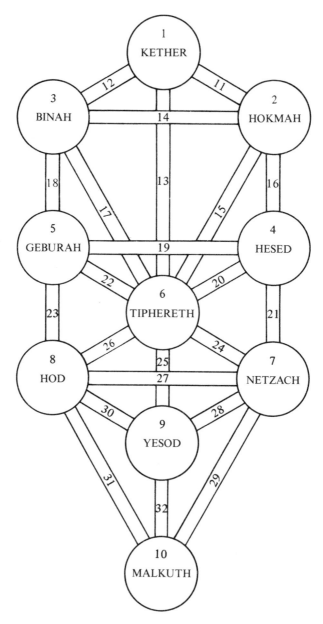

Sephirotic Tree

influence, the need for action in all its forms. That is the masculine temperament. And the feminine principle submits, receives, forms. These two poles, the masculine and feminine, the emissive and receptive, are necessary, so necessary that if one of the two is lacking, life cannot be present.

All beings, both male and female, must have both poles within themselves, in their inner psychic life. Otherwise, if they are receptive only, they are in for trouble, because they lack the will to do anything about what happens. A person who develops his sensitivity only, will be helpless to deal with the hostile forces he faces in life; if he is only emissive, only positive, that too makes for trouble because he will reject everything, he will receive nothing in the way of revelations. He will be strong and powerful, but only in a fight... for that is all the will knows... how to fight, how to annoy others, to upset things, to be blunt and brusque, thereby creating enemies. When he is receptive only, he runs no risk of making enemies because he says "amen" to everything and is therefore pushed aside and trodden on, and that is no good either! Whereas by possessing the two principles, solar and lunar (plus the earthly principle in order to be able to affect matter) then it is perfection, he lives in plenitude. In the esoteric Science, it is said that such a man is androgynous, that is, a complete being. In India, this state of perfection and plenitude is symbolized by the lingam.

So you see, to take and to receive... there is nothing wrong in taking on condition that what you receive is entirely pure, coming from the side of the moon that is turned toward the sun. If you know how to receive and give, you will gain health also, but if you only receive and never give, there will be a fermentation going on inside you and you will fall ill. If you give without receiving, you will soon exhaust your supply and be miserable. Wisdom lies in knowing how much of what you receive to give, when to give it and to whom... and there are many more things to tell you on this subject.

Before closing I would like to add this: when you are looking at someone, how can you tell what his thoughts are, whether the actions he is planning are good or bad? If his expression is gloomy, sinister, threatening, you can be sure he is contemplating some evil. Nature has so arranged things that a man bent on crime shows it in his face, his expression darkens and loses its radiance. If he is thinking of making efforts toward the good, of helping someone, his face is clear, the light shines through.

Now, what conclusion can we draw from these observations? What about the sun, why is its face always luminous and radiant? Because the sun is always thinking about the good, yes, and its light is directly proportional to the elevation of its thoughts, its selfless feelings, its love, its science. There! Had you ever thought of that? The sun shows us the way to be giving, enlightening, vivifying, loving. But people are so far from this way of thinking, this science, that no one believes it. "Oh, yes," they say. "Very poetic, very pretty..." but they won't believe it. That's it, everyone will agree, "How poetic, how lovely!", but no one will do anything.

How much there is still to reveal to you! For the moment none of this is clear because you are new to the science of symbols, but it will all come clear in time, dear brothers and sisters. Be patient!

Le Bonfin, 29 July, 1968

Chapter 4

Evil is Limited, Good is Limitless

The problem of the personality and the individuality is one that lasts all our life... not only this lifetime but all our earthly incarnations. We often believe we are being inspired from above when it is really only the personality trying to lure us into trouble. The individuality does its best to warn us of the consequences of listening to the personality, the inevitable consequences, but we prefer not to hear. That is the reason we should learn to observe ourselves all the time, ceaselessly inspecting each idea, each desire and impulse that occur to us in order to determine their nature and the nature of the consequences entailed. Unfortunately most people prefer to let themselves go along with the personality instead of avoiding its advice, with the result that sooner or later they suffer, sooner or later they are faced with remorse and disillusionment. If you could see into people's hearts and hear their confessions, as I do, you would be appalled. I often am. The things I hear make me realize how ignorant most human beings are concerning their two natures and the respective manifestations of each.

A man thinks for instance, that he is the one to benefit when he satisfies his lower desires, but it is not so. It is other entities, whichever ones he has chosen to work for who benefit, but he does not see this until later, too late, when he has lost all his joy, his strength, his inspiration. The time comes when he realizes that he has spent his whole life working not for himself but for the personality and its friends, myriad invisible entities that surround him without his knowing it, feeding on him as he gratifies their wishes. Then also comes the realization that he has left the immortal part of his being, his higher Self, to starve to death. Many entities feast on humans, too many to describe here. We have inherited them along with other things from the past and we drag them around with us as generations have done before us. If we could only see these entities, we would stop letting them take advantage of us and start working for beneficent creatures, the ones who would, on the contrary, increase our wealth and power, our real heritage.

Anyone who learns to observe himself discovers that after giving in to his desires, he feels robbed of all strength, lucidity, peace of mind... proving that others, not we, are the ones to benefit from our lower actions. If we were clairvoyant we would see the millions and millions of entities who use us as they will, robbing us of our energy. I met a writer once who had an exaggerated idea of his importance as a result of having had a novel published, and I touched on this question of man and his two natures, divine and animal. When he heard me say that creatures in the invisible world treat humans the way humans have always treated animals, he became highly indignant; "What nonsense!" he cried. "I don't believe a word of it". I stared at him thinking that for a writer of note he was not terribly perspicacious, for what I had said was true. Men consider it perfectly normal to make slaves of animals, to skin them and sell their furs, kill them and sell their meat, without asking the animal's opinion! If they did, animals would have a lot to say on the subject of human cruelty and injustice. Is it then so incredible for other creatures to treat us the same way? Is it not logical? We are thrown a scrap to eat, we are used to labour in the fields, and in the end we are cut up into ham and sausages for supper! If we could see into the invisible world we would realize that all the nations, peoples and tribes of the world's history are represented, some of whom have little or no regard for humans and use them as we use animals... they drive them too hard, they sell them and skin them and kill them and devour them. This will one day be a recognized fact... in the meanwhile you have the privilege of being the first to know.

Now do you see how important it is to observe yourself, to analyze your thoughts, desires, hopes and plans for the future, inspecting them as a jeweller sorts out the gems that are flawless from those of no value? When it comes to things like food, clothing, objets d'art, fruit or plants or flowers, men know a great deal about the different degrees of quality, and they admire qualities in people and things which are superior, they make a point of distinguishing between superior and inferior. It is not that some people prefer poverty to wealth, ugly women to pretty ones, rags to finery, no, what I am saying is that the Initiates base their knowledge on observation and they all agree that superior degrees, higher worlds exist, and that there is no limit to good, to excellence, we can always go higher. You cannot say about someone's intelligence, "Ah, he has reached the height of intelligence, he is the wisest", for there is no limit to wisdom or beauty, or goodness, or love. They are limitless and can be increased, augmented and perfected all the way to Infinity.

What about the opposite, things such as illness, misery, ugliness? Are inferior things also limitless and capable of expanding all the way to Infinity? No, there is a limit to evil, proven in physics by the fact that heat rises from 0°C. to infinity, whereas cold cannot go lower than 273° below. The frozen particles block each other and pile up; when movement stops, the limit has been reached. Heat does the contrary, it dilates and expands the body, stirs the particles into movement and pushes back the limits of space. Space is infinite, it cannot be limited. We believe ourselves limited because we have never tried to go beyond our own experience, we *think* we are limited in space but we are wrong: above there is no limit.

That is what led me to the conclusion that evil is limited in both time and space. God did not intend evil to endure. He did not endow it with lasting power as He did good: the power of good is unlimited. That is the difference between good and evil, the only real difference. People believe them to be equally strong but they are not. The forces of Hell are not equal to the forces of Heaven. Therefore, in electing to go toward the positive pole, you enter the realm of unlimited time and space, Infinity and Eternity, God Himself.

Morally therefore, we could say that those who choose the downward path, the personality and its weaknesses, disorder and chaos, are in fact choosing death and destruction. Little by little they are forced to disappear, they die because they are too tightly jammed together where there is no room: at the bottom of an inverted cone... where they can neither move nor breathe. The way of the individuality, the ascending path, gives them more and more freedom as they climb into the vast regions of space. All truly intelligent people choose that path, in order to avoid colliding with others and creating situations in which they are wounded and wound others. The alternative is to climb high into unlimited space and...freedom. The path leading downward limits one more and more until there is no alternative to destroying one another in the fight for space which is so vital.

If you try living for a day or two in complete chaos, you will see that one by one all your luminous, positive and beneficent forces leave you, while other, negative presences take their place: you find yourself in the dark, weighed down, uninspired and rigid. When you feel that way, you can be sure (it is a warning) that the space around you is shrinking.

You know what it is like to be caught in a crowd in the

street, the theatre, the underground, you feel trapped, afraid of suffocating to death... and when at last you emerge onto the street, you heave a sigh of relief at being able to breathe freely again. Well, the same thing happens in the spiritual world. It is strange that these experiences are common to all of us yet we never seem to learn from them, nor act upon them. If you have a feeling you are trapped, pushed into a corner, you must say to yourself immediately, "Oh! I must have taken a wrong turning, I have lost my way! Now I will do everything to get back on the right path...." We all become lost now and then for a few hours... or days, or years... yes, everyone. But how many realize that this language can be interpreted and interpreted correctly? The language itself is clear but there has been no one to explain it to you, parents and professors do not give instruction of that kind. In the Teaching of the Great Universal White Brotherhood, you are shown the Truth in order to help you to create your future, the future you have decided upon. Analyze yourself: how often do you feel (not by chance) that, although you may not be trapped in a crowd physically, you may be in your own room at the time, and yet you are assailed by some unknown force that torments you, something dark and strange that tries to crush you and rob you of your strength. It is because you have let yourself slide downward, backward, and now you must get back on the right path.

If we are so ignorant as to be deaf to all warnings, if we are blind to our situation, it will get worse; as in quicksand, the more we try to extricate ourselves by struggling the more firmly we become embedded: our feet find no firm ground. Do you find yourself in a swamp infested with mosquitoes and hornets and other tormentors? How can you expect them not to sting and bite, you are on their territory! You shouldn't be in those regions anyway, why did you wander so far afield? Those creatures have every right to resent your presence, it is their kingdom and they belong there, whereas you do not. You have taken a wrong turning. Now you must get back on the right path by obeying the laws of Nature and then the things that torment you, your difficulties and disillusionment, will vanish... or at least become insignificant enough to be handled with ease.

So many more things to say on the subject of the personality, but the most important of all is to remember that the personality is closely linked to the creatures of the underground. Although other creatures exist in those regions also who are devoted workers, devas, most inhabitants of the lower world are not very far along in their evolution, still in the dark, still egotistical and obedient to the personality. Whoever listens to the advice of the personality is vulnerable to the pernicious influence of the creatures, and in time becomes bound and limited... in chains.

I would like to add at this point that it is not necessary to get rid of the personality completely, to annihilate it... no, that is not the way to be free of it. The personality is there for an important reason: it provides the matter we must have to do the task we are called upon to perform, here on earth. Like a rich old granny who holds the purse strings, the personality has a store of treasures, a fortune in the bank, but it cannot be trusted as far as advice is concerned, it is too short-sighted, too self-centred, too selfish... but if you put it to work at what it does well, then it will accomplish wonders for you, as no one else can. It holds the key to the safe, to the larder, the cupboards, and you must not kill it! It will be useful to you if you make it listen to you, if you control it, rather than the contrary. People think they are masters of themselves when actually it is the personality which is in control of them! I see that situation so often that I have made a study of it at first hand... where do you think I learn these things I talk to you about if not by analyzing myself, observing myself? I notice the smallest details. It is not essential actually to experience things, to run the risk of losing oneself in order to study the effect of certain experiences. Why do people rush to investigate all that is dangerous, negative, corrupt and immoral, instead of examining them from a distance without becoming involved and damaging themselves? You needn't be a victim to understand something and draw conclusions! People, especially young people, insist on experimenting everything to the very bottom, the depths, not once but a hundred times, totally oblivious of the fact that they are exhausting their resources and will have nothing left when they are called upon to undergo higher experiences. If you think you can decide to turn to the light and learn about the luminous celestial spheres and taste the splendours of spiritual life after wallowing in filth all your life, you are mistaken. If you have used all your resources you will not have the slightest chance of rising above the conditions you yourself have created.

Once you have wasted all your energies on experiments in the lower world you cannot expect to be allowed to have divine experiences in the higher world. To think you are free to do as you like once you are morally bankrupt, shows how little you know of the Science of Life. Without the essential purity and freshness, energy and suppleness and intensity, you cannot open the door to Heaven. You think that even when you are dirty and corrupt Heaven will open its doors to you? How ignorant of you! No one ever has, only those few who barely skimmed through the lower world, refusing to sink into the mire, and who climbed back up as soon as they could to make amends for their mistakes and re-establish the vital current between them and the divine world, are able to have divine experiences. If what you want is the lower kind of experiences, then I say, "Go ahead! But I will be curious to see the results, what you will do afterwards." Nothing, that is all that will be left for you, it will be all over as far as you are concerned.

Good cannot be limited, good is limitless, which is why I think one of the best definitions of God is, "Wherever there are no limits, where Infinity and Eternity and Immortality exist, that is where God is." Nothing that is limited by time and space is God, nor can it represent Him. Like heat, good has no limit to how far it can rise, millions and millions of degrees, who knows how far. Last year at the World Fair in Montreal I saw the most extraordinary instrument in the Russian Pavilion designed to heat chemical and organic matter (and also plasma) to many millions of degrees.

If scientists had a correct vision of things they would see that Nature long ago established a scale of values with higher and lower qualities, which proves that there are two possible directions for all things to go in, one which is limitless and goes on all the way to Infinity and another which is evil and is limited. If you have never come to any conclusion concerning heat and cold it is because you do not know how to read Nature's Book, you don't understand her language.

Everything in the world below is a reflection of something in the world above. If you reason properly... yes, but your brain cannot reason properly, it is warped and destroyed by what you have learned, the habits you have acquired and inherited from your family, from the world, from the education you were given. It is the same for me, except that I have been working for a long time to get rid of all the old distorted ways and, if you do the same and drop old ways in favour of new ones, you will then see things as I do and make the same discoveries. You cannot deny the truth of these things I reveal to you every day, for they have existed since the beginning of time. People do not realize they exist because they are deformed by present-day culture. We must, therefore, either become detached from it or else resign ourselves to going through an entire lifetime without making any discoveries at all. As for me, I swim in the great truths I have discovered and which I cannot yet reveal to you, for until you have detached yourself from all the false concepts you inherited you will be able neither to understand nor accept them.

Imagine if you will a piece of paper on which there are two drawings, one sketched in red ink and one in green ink. You are handed a pair of glasses with which to examine them. If the glasses are red, you will not be able to perceive the red drawing, because red on red does not show up... and you will see only the green drawing which may be quite ordinary, even ugly. Now someone hands you spectacles that are green, and you are able to perceive the red drawing... whereupon you cry, "Well, but that is not the way it's always been... I am not used to this!" It has always been "that way," but the spectacles you were given to wear kept you from seeing the reality. For instance, when you look at Nature, the stars, trees, mountains and so on, what you see is an outline, is it not? Well, there is another outline behind the one you see. The spectacles I am offering you will permit you to see the second outline and more besides. What if you think you will no longer see the outline you are used to... so much the better, you have seen enough of that one. Nature has drawn two outlines, and it is up to you to find the hidden one. (actually there are many more than two, but let us say two for the sake of clarity).

Is this clear? Once you know there are two directions within you, you will try to avoid taking the path that leads downward, for it ends in a cul-de-sac, a dead-end, comparable to absolute zero (which in fact has never been attained) where all the molecules become completely immobilized. But if you follow the upward path you will have access to the Infinite – and what treasures you will find along the whole length of that path!

For the most part men live in slow motion: they make sure they have a source of livelihood, they start a family, and that is enough for them, they sit back, contented. But it is very little! When someone disturbs their complacency with new ideas (such as these), all they find to say is, "This fellow is a real nuisance! We are fine as we are, let him mind his own business!" Yes, men are accustomed to their way of life and have been for centuries... and I come along to bother them, stirring them up and trying to inculcate new ideas into their lives, hoping to make them a little discontented with what they have! No wonder they object, no wonder they are dismayed, even the brothers and sisters don't like to be hustled and disturbed! I cannot help it. This is my mission, given to me to carry out no matter what, whether it pleases people or not. My purpose, my duty is *not* to leave you alone! And what you complain about so bitterly makes the divine world rejoice, they applaud when they see my efforts! I understand that you do not appreciate my methods, but nevertheless the world above finds them salutary as far as your evolution is concerned.

Le Bonfin, 3 September, 1968

Chapter 5

Eternal Happiness

I

"It is in the silence that we hear the Voice of God. If this is the Voice we listen to and allow to guide us, then we are treading the path that leads to eternal happiness...."

This Thought-for-the day requires a little explaining. It is no coincidence that the words "happiness" and "Eternity" are mentioned in connection with the Voice of Silence.

Let us look at the way most people think of happiness. "Ah!", they say. "I am so happy! Never have I known such happiness! I would be happy if I had this or that...". Their concept of happiness is that of a passing sensation, something that affords a few moments of satisfaction and contentment such as sitting down to a great feast, or going on a long voyage, or kissing a pretty girl... which, agreeable as those things may be, are no more than sensations produced by the five senses. That is not real happiness, dear brothers and sisters. We are in error when we think that a wife and home will make us happy eternally... and the same for fame and glory, knowledge and learning, beauty, success, etc.... Thousands of years of world history have proved to us that happiness is not to be found in those things, or at best only temporarily. To understand happiness as the Sages, the Initiates, understand it requires a thorough knowledge of the structure of both man and the universe he is part of. Without it, you will never know real happiness.

You all remember the story of the Yogi and Alexander the Great when he was in India. Alexander heard tell of this most exceptional being, a Yogi who lived in dire poverty and at the same time in utmost bliss and felicity, and he went to see him. The Yogi was seated in meditation, an expression of beatitude on his face. Alexander announced his intention of taking the Yogi to live with him in the royal palace in Macedonia... where he would have all honours and riches. The Sage smiled in compassion for Alexander, and replied that he had no need for such a life and was very well where he was. "What!" cried the great conqueror in anger. "Refuse my offer? Don't you know, poor wretch, that one word from me and you will lose your life?" The Sage smiled again, "You cannot do anything to me, death has no hold over me... I have conquered it. Not I, but you are the one to be pitied, all your possessions must be a terrible burden for you... how unhappy you must be in your heart!" For the first time in his life, Alexander the Great, fresh from military triumphs over all the armies of the world, was conquered by a man without a single weapon! He went away crestfallen. How had the Yogi obtained happiness? You will see.

They say that primitive, simple people are happy, that it is only civilized man who has such great difficulty finding happiness, because in developing his mind and sensitivity he becomes more and more vulnerable to events and the material and psychological conditions he lives in. Some people would rather be primitive than civilized, since one can be happier that way! In that case, why not keep going further back, and live like an animal? Animals are happy. Or go still lower down amongst the plants which do not suffer, supposedly. In that case, to be a stone must be the best condition of all, for they do not even feel! What logic!

Happiness as people usually understand it is a state of mind which lasts momentarily. For a few minutes they can say, "Ah! I am happy, life is good!" But a few minutes later they no longer feel that way... why? Because that happiness is based on people or things that are changeable, ephemeral, temporary. Happiness exists, yes, but only when it is the result of something unalterable. Initiates have discovered the regions in which real happiness exists and that is why they devote their lives to thinking, feeling, loving, working, doing things in such a way that will permit them to live forever in those regions where happiness reigns without cease.

The reason happiness is hard to obtain, and once obtained so difficult to hold on to, is that it takes superior qualities, above all purity, for only then will happiness be lasting. How can you expect to have a lasting happiness when you are full of hatred, jealousy, egoism, greed? You may be happy momentarily because you have gained by taking advantage of someone, but very soon afterwards you will find you have difficulty sleeping, worry and anxiety take the place of peace and contentment... is that happiness? Not real happiness in any event!

Real happiness lasts. You say that nothing lasts, life is nothing but change, success alternates with failure, abundance with poverty, peace with war, health with illness... but is there no way of avoiding that state of affairs? The happiness I am talking about exists in spite of those changes: war can break out, you may lose your entire fortune overnight, but still you are happy! Why? Because your consciousness is above the level on which events take place, you know the truth behind each difficulty, each trial and hardship, and that truth comforts you and leaves you in peace no matter what happens externally. When you are above the level of events you know how to reason, how to put things in their proper place. You may have things stolen from you, you may be tormented in various ways, but none of those things make you unhappy... since you know they are only temporary and that you yourself are immortal! Nothing can touch you or affect you under those conditions! Events which terrify most people merely make you smile.

Real happiness can only be found far above, in the same region where harmony, purity and love are. Happiness is inherent to all of us, but when we lead a superficial life, on the periphery of ourselves and our lives, then we are not conscious of how happy we are, for there is nothing but change and illusion at the periphery. A few joyous particles may visit us for a minute or two, but suffering quickly replaces them, as though to punish us for stealing a few moments of happiness! No one gets by without paying dearly for stolen happiness.

You must have no illusions about happiness: it can be yours, yes, but only on condition that you have worked to obtain light, purity and stability, for happiness is composed of those elements. Like peace: I have told you, peace is not one element, it is a synthesis, and the same is true for happiness. Happiness is a synthesis of a great many elements of which the most important is stability. When a man has been deceived and disappointed by all that is temporary and illusory in the world, he turns to the immutable, eternal Spirit, God, and... finds happiness! Once found, he will always have it, no one and nothing can ever make him unhappy again. Regardless of his situation, whether he is rich or poor, famous or scorned, with or without a wife and children, he is above change, above human foibles, he lives in Eternity.

This language is not one that is easily understood. How can it be? Some nitwit says to a girl, "Darling, I will make you happy..." when the poor thing doesn't even know what happiness is himself, let alone how to make someone else happy! Or a young girl will whisper to the boy she "loves", "I will make you happy...." How, with all the imperfections, moods, jealousy and fits of temper both are subject to? How can they make anyone happy and live happily ever after as the fairy tales say... is that happiness to be relied upon? That kind brings a moment of joy now and then, like the brief respite felt by prisoners when they walk and breathe in the open air before returning to their cells. Or like toothache, it stops for a second but then begins again, twice as painful.

To be happy, my dear brothers and sisters, you must have a fixed point, a centre to hold on to permanently so that you will never again lose your balance; in physics it is called an equalizer, a pendulum that swings from left to right but always returns to centre because it has a fixed point. We must find that fixed point in ourselves, and always return to it, the fixed point, our equilibrium. If we do not know how to think about things, how to feel and act, then we are uneasy and afraid, at the mercy of every wind that blows. As I told you yesterday, happiness is no more than a state of consciousness, a way of thinking, understanding, feeling, behaving, it is an attitude. To have the right attitude requires study and the knowledge of a science which you put into practice. Happiness is not something you obtain without spiritual work, discipline, effort. It is a synthesis. If you understand things as they really are, if you feel correctly, then you probably will act as you should, which is what will make you happy! To reach that point however, it is advisable to enrol in an Initiatic School, to learn how to control your mind, heart and will, which is the same as saying to learn how to be happy! Otherwise you will be no happier than other people.

Happiness is not to be bought at the market like a pound of cheese or a loaf of bread. It requires discipline through which little by little you raise your love, your understanding as high as Heaven, where you can then draw on a boundless ocean of love and felicity. But first you must acquire stability, steadfastness, and be able to say with the Initiates of Ancient Egypt, "I am steadfast, son of steadfast, conceived and born in the region of steadfastness". But as long as you are hesitant and changeable, you needn't look for happiness! You'll taste a few minutes of joy and *think* you are happy, but that happiness goes away almost at once and you are left with your tears... to weep and weep.

That is the way it is with love. You know the song, "The joys of love last but a moment... sorrows of love last a whole lifetime." It should be the reverse! Once you find it and grasp it no one should be able to take it from you, it should last forever, like happiness. Ah, but love! People search and search for love and never find it because they stop short at the lower level, effervescence, fireworks that are quickly extinguished, leaving them once again in the darkness.

Happiness is not the destiny of stones and rocks, nor of plants, nor of animals... nor of humans, it is not given to man to be happy, nor even to supermen who may do glorious things and accomplish wonders, but who are not happy. Happiness begins on a higher level with the kingdom of the Angels, whose consciousness is beyond all impurity, beyond darkness. Angels do not know suffering, only happiness! But man suffers, and supermen suffer... more than ordinary men, for they have greater sensitivity.

What makes primitive men happier than civilized men? The fact that they live close to Nature and are content with little. Peasants are content to live with their wives and children and animals all together in the same hut... it may not be sanitary, nor aesthetic, it may smell a bit, but no matter, they are happy! The more civilized a man becomes, the more he studies and learns, apparently the more miserable he is. Why? Because the learning and experience he acquires make him more difficult, more demanding and exacting, more self-centred, his needs increase along with his desires, and everything becomes more complicated. He can't get on with others, even with his own family. The cause of it all is education, today's standards of education, because they call for a highly developed personality with all its egoism; everyone pulls the

bed-clothes over to his side and that is what causes quarrel-
ling, separation, divorce, etc.... Primitive people have to put
up with each other, their demands are less and they take what
is offered without looking for more.

If the world's education were put in the hands of Sages,
Initiates, they would make sure that people knew what method to use in order to develop the higher nature, the individuality, and not the personality of youngsters and learn to be
generous, impartial, impersonal and unselfish, as well as to
know facts. Nowadays people are not really educated; the
personality, not the level of the mind, is stressed until each
person thinks himself the centre of the universe, with all
others there to serve him! How could they possibly live together in peace under those circumstances? The fault, the
blame belongs to the educators, the heads of schools and
universities who have oriented education in the wrong direction. Supposing I were given the task of directing education in
the world: I would guide it in an entirely different direction,
and the youth of the world would be completely changed...
simply in a matter of a few years.

Believe me when I say the personality has been overdeveloped; having made a very complete study of it, I know
what it is capable of. Psychologists know many things which
are no doubt useful, but to me they have overlooked the most
important, the essential question: that of how to handle the
problem of the personality and the individuality. Everyone
should be shown that it is these two natures in man, the lower
and the higher, and the way we behave as a result of obeying
one or the other, that create the consequences. I have given
my entire consideration to the subject of the personality and
the individuality, and now I have the knowledge, the key with
which to solve all the problems of life.

Many people prefer to stay in their own holes quietly and
thus avoid trouble. From time to time they feel the need to
meet others, to talk or dance with someone, but on the whole

they prefer to remain separate, closed off from other people, preoccupied with their own peace and tranquillity. Why? And others find that it is precisely in the collective life, in the midst of people that they find strength and inspiration and advancement. The first group is under the influence of Saturn, the second is influenced by Jupiter, two complementary planets. Saturn is sad, solitary, pessimistic, lonely as a hermit in his cave, asking nothing but to be left alone and undisturbed, especially where women are concerned. The Saturnian type has a particular loathing for women, perhaps his wife abandoned him, poor thing and, being vindictive and spiteful by nature, his grudge extends to all women. He still wears his wedding ring (the ring around Saturn) but he wants to be alone in his misery. Jupiter, on the other hand, is generous and smiling, full of love for everyone, wanting to be with others so as to distribute his riches, to share with others... for the Jovian type must give.

Whether I am Saturnian or Jovian I do not know, but when I analyze my reactions, I see that alone I am bored, somnolent, uninspired, dull... and when I am with the Brotherhood I am happy and stimulated... is that Jovian? Some people are uncomfortable in a collectivity, they feel put upon and hurry back to their holes as quickly as possible... which is very revealing of their characters!

That is not all there is to say on the subject of happiness of course, but the most important thing to remember is that you cannot be happy as long as your vision is narrow and limited; to be happy you must expand and grow, stretch all the way to infinity and embrace the vastness, the immensity of the universe.... Infinity! A man who prefers solitude cannot be happy, his selfish personality makes him shrink from space and the infinite. It is love that makes you expand and unfold until you embrace the whole universe and vibrate with it. Then you no longer meet with obstacles, then happiness comes to live with you for good. Hence, the way to happiness is through love, unlimited love, real love. Knowledge does not bring happiness, nor does science, nor philosophy... people who know a great deal are not really happy, whereas those who are ignorant but who have a heart full of love are happy... why? Because God placed happiness in the heart, not in the brain. Learning, science, prepare the way: they enlighten us and orient us in the right direction but they do not make us happy.

Too much learning and knowledge makes people unhappy. "Great wisdom brings great sorrow; much learning brings much pain", is a saying attributed to Solomon. People worry when they know too much, knowledge produces light and in the light you see all kinds of things better left in the dark! Happiness comes from the heart, from loving, therefore we must love, but love wisely. Love and wisdom are closely linked, wisdom shows love the light, and love shows wisdom how to be warm, for wisdom is cold. The person who has both love and wisdom has found Truth. Unless wisdom and love are together, Truth cannot exist. You may think you are doing the right thing, but if you are missing one or the other, you are not in the way of Truth. A medal that has love on one side and wisdom on the other, that is the medal of Truth.

All this is confusing to whoever is hearing it for the first time, or has not had sufficient preparation. It is not the way we were taught originally and it will take time to understand. If you are patient however, even if it sounds completely senseless, contradictory and not a philosophy worthy of the name... in time you will see how true it is. You have studying to do and experiences and suffering to go through before you can realize how absolutely true it is.

Let us return to the Thought I read you beginning with "It is in the silence that we hear the Voice of God...." As I said, silence is the preliminary condition in which all dissonance, discord and difficulty are eliminated. When the silence is completely harmonious, you discover that something speaks to you : this is what is called the Voice of God, the soft gentle voice that warns of impending danger and directs us on to the Path. When we do not hear it, it means we make too much noise, not only physically but in our thoughts and emotions (for they also make a noise). When we are really quiet, the Voice reveals to us that God alone is Eternal, God alone brings happiness, He alone explains what happiness is. The Voice of the Silence* is also the title of several books on the Eastern Wisdom. A Yogi who establishes peace and quiet within himself and stops all thought, hears the Voice of the Silence, the Voice of God. I too have heard it.

One way only exists to express the Divine Nature of God in all His Splendour and Glory : silence. You are not yet able to be silent inside, and so, for the moment, I substitute myself for the Voice and tell you in words what you would hear in the Silence if you could. If finally you do hear the Voice, in a fraction of a second all will be clear to you, everything I have just told you will all be revealed. The Silence, the Voice of God, happiness, Eternity, are all connected.

It is very difficult to reach the true state of happiness, much work is needed before you are allowed to enter the world of Divine Love.

> In Divine Love lies true happiness
> In Divine Wisdom lies true light
> In Divine Truth lies true freedom

These are formulas, as you see. Happiness does not come from without as people think now, from objects or possessions. So many bizarre ideas have been released in the world that people are at a loss, adrift in their ignorance. They think, "Ah, if I could only buy this, have that, I would be happy!",

---

* The Voice of the Silence is the translation (HPB) of the Sanskrit and means the Voice of *the* Silence, or God, and hence the Voice of God.

but happiness cannot be obtained that way, at least not happiness that lasts. How often we buy something we want and then are still unsatisfied, inwardly the void is still there. If you seek happiness inside, within your love and understanding, then even if you have no possessions at all, if you are stripped of everything others consider essential, you are happy and soar high above, because the reason for your happiness is within. It is not a bad idea to have something externally also... if you handed me a million or two, I would not be offended, I would not refuse... try it and see! But if you tell me that I will be happy as a result of the money you give me, no, not on your life, it is simply not so. I know I was extremely rich in other incarnations, and it never brought me happiness. Happiness does not come from outside.

I am showing you the way to happiness: you must look inside yourself, you must adopt a certain way of looking at things, of understanding and feeling, that is what brings happiness. If you do not believe me and do not wish to accompany me on this luminous Path, I will keep on alone. But I know I am not the only one to believe this way, many others in the world think as I do. And I have more and more hope that one day we will be numerous enough to improve the condition of humanity: this hope makes me happy!

I wish you true happiness, dear brothers and sisters. You will find it in the Light and Love of the Spirit.

Le Bonfin, 19 July, 1970

Eternal Happiness

II

I would like to add one more thing to what I was telling you yesterday about the personality: its most pronounced characteristic is its fear of silence. The personality much prefers noise, fighting and arguing, chaos; disharmony is where it feels at its best and happiest and where it can breathe freely. In the silence, it feels paralyzed. The silence keeps it from giving vent to its native arrogance, truculence, changing moods, conniving. Young people for instance love noise and agitation because the personality is strongest in young people, there is nothing to keep them from running after sensations and pleasure and emotions, no discernment on their part, and this leads them into mistakes of all kinds. The adult has learned by experience that pleasure does not bring what he is really looking for, and he accepts silence in order to meditate, to think back over the major events of his life and draw certain conclusions.

People in whom the personality is highly developed live in constant uproar. Today's idea of music, for instance, is no better than uproar! When I was in Japan at the World Exhibition at Osaka, I heard an orchestra that nearly drove me mad. I had the impression while listening to it that it tore the nervous system apart and destroyed the soul. Painters, musicians, sculptors, no longer know what to create, it would seem that art has been exhausted, and artists are at loose ends. In the past artists knew where to go for inspiration, high above amongst the forms, colours and melodies of Heaven, which they brought back down and reproduced in their work for others to revel in. Now artist have taken the path downward and bring back patterns and sounds from Hell, as if they were bent on driving mankind mad. Mankind is already a bit touched, but this modern music will be the end of it!

Few musicians take the trouble to study the way vibrations affect the listener. I spoke to you about the experiment made by Chladni, the great scientist, who put a little powder on a flat piece of metal which he then caused to vibrate with a bow: the vibratory waves create lines of force, attracting particles from the points where the vibration occurs (called "live" points), which are then rejected toward "dead" points, where there is no vibration. Geometric figures are formed by the "dead" points. I have noticed that this experiment also can be applied to the human body: there are "live" points, that is, centres which vibrate and project elements which go toward other "dead" points. The sounds we hear form geometric figures inside us. We may not see them, but the vibrations create infinitesimal particles which form figures. The way God created the world by the Word can be explained in the same way: the Sound uttered by the Creator made vibrations which formed the geometric structure of the world. You can also make an experiment by letting a ray of sun come into a room through a tiny hole in the ceiling: if you shake a little dust onto this ray whilst playing a musical instrument such as the violin, for instance, the sound will have an effect on the dust and cause it to form shapes, designs. This is a law of physics. And when you listen to contemporary music in all its cacophony, you deeply disturb the original order established within you by the Word of the Creator. Another order takes over, something chaotic, disharmonious creates a disorder within that reflects on you even to the expression of your face!

Music is not the only thing to produce this effect... have you ever noticed the way people's expressions change? What causes them to change so suddenly from a pleasant, smiling expression to a disagreeable, repulsive one? We have seen that sounds produce vibrations: a person's thoughts, feelings and energies also produce vibrations. Your inner state produces pleasant or unpleasant vibrations just as harmonious or disharmonious sounds do. It does not seem possible that man has not understood this law! If we make a point of studying it now and consciously, deliberately, introducing harmony within ourselves each and every day of our lives, we will then always have a beautiful, pleasant and harmonious expression on our faces!

Everything is a question of vibrations: the Initiates know that. They have studied the whole question of vibrations and what causes them, with the result that they do everything to make us see that a man can, by his inner life, not only shape his body exactly as he wishes it to be, but also the rest of the world in which he lives. Conditions are good or bad, his life is full of success or failure, happiness or unhappiness according to *himself,* he is the cause, consciously or unconsciously. This science is hard to believe, but nevertheless it is the truth! I have verified it.

These are a few words on the subject of music and harmony, and the disharmony which exists today and is being spread about the world. The word society uses for disharmony is anarchy; unfortunately there are too many candidates for anarchy in our world. If we knew what we are preparing for the future by this attitude! A disharmonious state... of course when I say "harmonious" and "disharmonious" you understand that I am talking about more than music. To me a disharmonious state is one of hostility, vileness, destructiveness, cruelty, the supremacy of the personality... and a state of harmony is one that is impersonal, generous, noble, pure, filled with love.

Why is the personality afraid of silence? Because during a silence conditions are not right for it, it cannot do what it wants to do. Silence is a door opening onto heavenly regions where the personality is uncomfortable with its purely selfish projects, its attitude of thinking that everything is owed to it, that it can get its way by fighting and clawing and biting. The silence does not tolerate that kind of thing, and the personality is forced to give in, to capitulate, which is not what it wants! At the slightest opposition, instead of advising calm, the personality says, "Fight back! Bite him! Exterminate him!" The advice of the personality is never anything but destructive, whereas the individuality says, "Wait a bit, hold on! Try praying for your adversary, send him some good thoughts. He may change and then you will have a good friend instead of an enemy. No need to worry about yourself, no one can destroy you, for you are eternal! Simply try to have a little more light, and give a little more love!" That is the individuality. But the personality is so loud, it uses big drums and blaring trumpets that make a noise night and day, insisting, demanding, exacting and, finally we give in, we think (stupidly), "Oh, very well! Better do what it says, maybe it's right." The voice of the individuality is unobtrusive, so gentle it can barely be heard... which is why we are always hearing the personality.

I envy the personality one thing only: its indefatigability. The rest is dreadful, but it has that one quality... like criminals, they are indefatigable, inexhaustible, because their diabolical projects keep urging them on. Whereas the good, the kind, the amiable people in the world are always tired! They have no enthusiasm, nothing to spur them on to steal, kill, get their revenge... and so they sit back, content with being lukewarm. One day I will show them how far they are from the real work. They haven't even begun yet. When they realize how much there is to be done, then they too will become indefatigable, but first they must seek, they must accept the High Ideal, and not be content with the tiny ideal they inherited... earning their living, fulfilling their marital duties, bringing up their children and feeding the hens, a miserable, insignificant, uninspiring ideal! Now they must decide to go much further, to undertake a tremendous, endless, vitally important task! In a divine School, you are shown how many magnificent things there are to do which you do not now even suspect exist. Once you have a High Ideal.... Ah yes! You must have a High Ideal.

The personality hates silence, silence strangles it. The meditation, concentration, praying practised in an Initiatic School is for the express purpose of reducing the personality, and allowing the individuality, the spirit, to come to the fore. In the current magazines and literature you are urged to content the personality above all: "Take this, take that, you will be happy!" Always contenting the personality, with no idea of doing something for the benefit of the individuality, the divine side, always everything for the personality's pleasure and comfort. When people are satisfied to the point of saturation they become dreadful... their personality has been overfed! Today's films, novels and plays are all aimed at the personality, and for the individuality, the mind and spirit, there is nothing at all, or practically nothing. No wonder things don't work out. But the personality is full to overflowing, it vomits and soils everything it touches... which is not surprising seeing that it is what it is.

What is silence, why does the personality hate it so? In a person who is still very young everything is in constant turmoil, tornadoes, whirlwinds, noise and excitement burst forth from him at all times. It is not possible for the divine side to manifest as long as the person is in such an upheaval. With the years those conditions calm down and make way for silence: the individuality then appears; before the calm its qualities could not get through. It is the same with Nature when, before the end of winter, a day of sunshine makes plants or vegetables bloom before their time and then, when the frost comes in the night, they die. The conditions were crippling. Well, that also happens to human lives when they are disturbed by storms and tempests, people cannot hear the inner voice, the voice of Wisdom, the voice of the Angels. Passions must be stilled for the good qualities to blossom forth in safety.

Still another argument, taken this time from the field of geology. In the beginning, the earth was in continuous fusion, life as we know it was not possible. Even after millions of years, when the earth had cooled and its crust had settled enough to allow plants and animals to appear, still periodically all life would be annihilated by earthquakes and volcanic eruptions. When the cataclysms finally ceased and atmospheric conditions became stable enough to allow the earth's crust to solidify, plants slowly appeared and attached themselves to the ground, then animals appeared and moved about on its surface, and at last man made his appearance. This image comes into my mind when I see someone who is in the state the earth was in at the start, and I say to him, "My friend, how do you expect the luminous spirits to come and install themselves in you when you are in such a turmoil? They know better than to risk being engulfed in the abyss and would rather wait until you have calmed down a bit."

How clear this is, how limpid! Now go ahead and try to do everything to create stillness in your inner life. In silence, if your emotions are under control, your divine side will blossom out at last in the form of beauty, light, virtues. But, before you have made room in your inner life for silence, no use hoping for the divine Spirit to manifest itself... would it be so stupid as to live where the ground is continually threatened by earthquakes?

A genius, whether artistic, mathematical, literary, scientific or spiritual... how does he become like that, where does his genius come from? You say, "It is a gift". Yes, but a gift of what? It is the gift of an entity sent from above to live in a particular person and manifest its genius through him. No psychologist will agree, naturally, that talent and genius are the result of an entity living inside one! But the proof is the fact that this gift sometimes disappears. It happens to many people who live a disorderly, chaotic life, they lose the touch of genius they originally manifested. Now, would you like to attract the higher entities and have them bring you gifts of all kinds? Then create harmony and silence within you: those are the conditions. The entities are waiting for such a signal, longing to see that someone has known enough and thought enough about them to create peace and order within... whereupon with what joy they rush to help him, and others through him! You did not know, did you, that once a higher entity decides to help you, you can then help others because of him? Why not work for that rather than being always in the same crippling state of agitation?

These are truths which, if men believed them, would change their lives to the extent of making them permanently happy... with the true happiness I described to you yesterday. The astonishing thing is, you become happy for no reason, you cannot say why, but life is wonderful, everything... eating, talking, breathing, becomes a source of delight to you! Yet nothing new has been added, no sudden bequest has brought you a fortune, no stroke of luck, no pretty woman has come into your life... but you are happy, happier than you have ever been in your life. Something from above is present all the time inside you that makes no demands, no sign of dependence or claims... like a stream of water flowing from Heaven! Earthly joys always involve possessions, houses, money, awards or decorations, women or children. As long as people cannot have what they want they are "unhappy", "happiness" depends on what they "have". When they lose these possessions, whatever they are, then what misery! What grief! No, true happiness does not depend on objects or possessions, it comes from above, an astonishing state of mind that makes you rejoice permanently, for no reason. That is happiness. If you say, "Ah! I would be happy if I had this or that...", you are relating happiness to possessions; but no matter how happy you think you would be, how long would it last? Happiness is short-lived unless it is true happiness, the happiness that flows abundantly from above.

True happiness is like air: can you go to market and buy a kilo or two of air? No, air is all around us, we are plunged in air as in an ocean and we breathe it unconsciously. Everything else, water, food, money has to be found or bought, but air is free, air simply exists of itself, as does light. There is no greater joy than breathing! If you don't believe it, try holding your breath for a few seconds, you will see. Well, it is the same for happiness, happiness is like the air you breathe. Once you are in the ocean of silence and harmony, you no longer need to look for happiness, you are immersed in it! The soul breathes in... out. You inhale... exhale. I know you have never thought about the way you breathe, you are used to going to get what you are conscious of needing one piece at a time, and this lets you experience a minute of pleasure, of joy. But with air, you breathe all the time, freely, even when you are asleep.

Breathing exists to show us that tangible things such as money, objects, possessions, etc... are not to be compared with the subtle, intangible etheric world of which we are a part. Once you are conscious of being immersed in this etheric spiritual world, you breathe more consciously, and this adds enormously to your happiness.

And so, dear brothers and sisters, make the effort to breathe consciously, and above all not to doubt, never to lose faith. I have never doubted. Since the first day of Creation I have always believed in the splendour of the Truth, and have continued against all odds, ridicule, hostility, injustice, misfortune, illness, poverty, hardship... and I do not regret doing so. Why don't you start today to believe, what are you waiting for? You don't know what you are missing! Hurry up, it is quite easy, believe me!

The thing to retain from today's talk is that you will not be any happier from letting yourself be guided by the personality. Those who are always pulling the bed-clothes over to their side, trying to be the centre of the universe and expecting the whole world to revolve around them, to serve and obey and bow down to them as before royalty... are paving the way for disillusion and suffering. We must be willing to serve. That is when the personality disappears. But is anyone a willing servant? People admire those who get ahead and succeed, even at the expense of others. "Ah!" they say. "That man is truly intelligent, he knows how to get on in life!" No, that is not true intelligence, people confound intelligence with craftiness and it is far from *intelligent* on their part to do so!

If this Teaching spreads in time (as it will) every single person in the world will be given the place he deserves. Those who have succeeded dishonestly and who are sitting at the top, proud as peacocks, will topple from their pedestals. Yes, this Teaching is capable of making anyone fall who has gained a high position by dishonest means, or rather, such people will in fact make themselves fall, before these Truths they will become disgusted with themselves, their self-distaste will force them to capitulate. Dear brothers and sisters, you may not believe me, but sooner or later, it will be so, for Heaven has so decided. Everything as we know it will be turned upside down. When this Teaching becomes known it will upset human consciences, for the Teaching, dear brothers and sisters, is a most fearful, awe-inspiring, invincible Light.

Do not have faith in the promises of the personality. It will

try to convince you to do all kinds of things under the guise
of making you happy and you will be tempted to go along. Of
course there will be a few happy moments, but later, when the
whole platform collapses, where will you be? The personality
excels at getting what it wants, it is an expert at music, poetry,
the dance, everything beautiful... it has great charm and expresses itself extraordinarily well. But its real aim is to devour
you. It knows how to look into your eyes with love, it can
soften you and pamper you, yes, but what it wants in the end
is to trap you, to enslave you, to devour you. Why does it
make itself beautiful? The better to get you! You didn't know
that, did you? The individuality is even more beautiful, musical, poetic and learned, not in order to bind you hand and
foot and treat you like a servant, but to free you, to rejuvenate
you in all your youth and beauty. The motive, the intention,
that is the important thing! Until you know what a man's in-
tent is you cannot judge him. If he offers a beautiful jewel to a
young girl, what is the reason behind this beautiful gift, why
does he offer it to her, what makes him do something in appearance so generous, so wonderful? His intention is to take
advantage of her.

The personality is far from stupid, on the contrary, it is
learned and clever, skilful enough to make the stars fall down
from Heaven to convince you not to go on doing the good action you have started, to cease your spiritual efforts. And you
will be convinced, because the personality is not alone, it has
the aid of a host of scholars and artists and brilliant people to
throw gold-dust in your eyes... a whole world crawls around
inside the personality! And I haven't even begun to tell you
all that there is to say about it.

Yes, dear brothers and sisters, rejoice at the fact that you
have this light, this discernment that helps you to dominate
and control your personality, for rich and gifted and capable
as it is, it must be dominated and controlled.

Le Bonfin, 20 July, 1970

Chapter 6

Fermentation

You must not become used to waiting for everything to come from me, dear brothers and sisters... for me to make you happy, for me to smile at you with love. Waiting for the other person to make the first step, to approach you is not the way to build your future. That is the way the world does it; people expect and demand attention and think everyone including Almighty God is there to satisfy their desires, to wait upon them hand and foot... and they are resentful when it is not the case! This attitude shows that they are controlled by the personality. The most pronounced feature of the personality is to demand, without ever wondering whether people are in a position to give what you are asking.

It is rare for anyone to put himself in anyone else's place; very few do it. This is a school for developing impersonality, where you learn to acquire the habit of expecting nothing from others, of taking the first step towards them, and in that way you become stronger. If you wait... well, you may wait forever, because others are busy with their own problems. You have no idea how selfish, stingy and self-centred people are, but... if you analyse yourself you will see that you are that way too!

It may be that Heaven thinks my children are very badly brought up: in the past a teacher, a Master did not behave as

I do, they were always solemn, impassive, chary with their smiles and praise, whereas my attitude is so different it will never bring results! I know I inherited this attitude from my mother, who always took the first step toward others. If I were to stop smiling at you and giving you loving looks and chatting with you... would it ever occur to you to send me a little encouragement now and then, to strengthen and help me? Not that I ask anything from you, no, but I can see how absorbed you are in your own affairs, your own interests. If you would try to become a little more generous in your thinking, a little more impersonal and unselfish, it would give me the greatest joy... and even more joy to those who sent you, who watch over you. We all have a guardian Angel who watches over us, the one who sent us here originally and who watches us to see what we do. "Oh", he says. "You are waiting to be washed and fed and encouraged by someone else? When are you going to start doing something for others?" God knows when. Why not make a try? Otherwise I am afraid the Higher Beings may close the door and you won't receive from them any more. I don't like it when Heaven shuts itself off, I like the water to keep flowing and flowing!

Do not think I am complaining. Why would I complain now after all these years, just when there is a little progress? You know parents, fathers and mothers, they are never satisfied with a little progress, they want their child to become perfect as quickly as possible!

The truth is, I tell you frankly, it is men's selfishness and narrowness that keeps such things as joy and inspiration and happiness from visiting them more often. They drag themselves around looking miserable, downcast and sad because they have not learned how to open themselves and make something flow out of their hearts for the good of all Creation! That is what they must learn. No matter how many lectures I give on this subject, you forget... when will you learn to think, to ask yourself, "What can I do for my

brothers and sisters, how can I improve their situation?"
That kind of thought makes you grow, instead of always looking for someone to help *you*, to give you money or console
you. Analyse yourself, you will see : girls, boys, young and old
people, all of them think only about using people in the hopes
of getting something from them. You say, "Well, but you too
are trying to get the whole world to work for you!" Yes, that's
true, I do want all mankind to work for the realization of
peace and happiness on earth.

It's true, I do want to gain all the workmen I can, but not
to cast a spell on them, not to subjugate or harm them. If you
do not believe me, go ahead and verify it! But at the same
time verify your own behaviour, for I can tell that you came
here to learn, to gather a store of riches, to strengthen yourself, rather than to help me in my work for humanity. The
proof is, when you have had your fill, when you have taken
all you want including even a boy or a girl to marry... instead
of going on with the work, you abandon me. This shows that
you came for yourself only. If you came with the idea of the
Teaching, even if you learned everything there was to learn
and understood everything there was to understand, and also
succeeded in finding a lovely young girl to marry... you would
stay in the Brotherhood in order to help with the realization
of its work. Even if there were nothing more to learn you
would go on working. The time comes when the brain has no
more desire to absorb, but the work is never ending. You can
only study for a few years, but you can go on working until
you drop!

If you come for yourself only, looking for someone to love
you or for someone's friendship, or even only to learn and to
accelerate your spiritual evolution, it proves that your ideal
was not terribly glorious or sublime. You are seeking your
own happiness, your own advantage, your own salvation. Is
that such a wonderful goal? We should stop trying to save our
own soul. What is so important, so valuable, so precious

about our soul compared with the immensity of all Creation? If people would stop thinking about their own salvation and think about saving someone else's soul, then they themselves might be saved! If they are busy saving their own soul, no one else matters, they remain isolated from the rest of the world, thinking of no one else, doing nothing for anyone else... because they are busy thinking of their own soul! That makes no sense, we must put all that aside now. The day humans get rid of the idea that they must watch out for the best for themselves, the world will be transformed. That idea is keeping the Kingdom of God from installing itself on earth!

We must learn to forget ourselves a little. How? By listening to the individuality. That is a gigantic work which very few people know: to forget oneself and the lower nature, and to think about and summon the divine nature. As soon as we think about our higher Self, which is great, vast, universal, everything falls into place, whereas if we go on thinking about nothing but our little *me*, limited, selfish, self-centred, nothing works. That is why there are so many unhappy Christians, they think themselves damned, they even commit suicide, at the thought that they have not succeeded in saving their souls! They don't realize that what they call their soul is only their personality... that is the mistake. The personality can never be saved, it will always be what it is....

I have come to demolish your confidence, that's what I am here for. I have brought big hammers with me to break down the hopes you have placed in your personality, for whatever you do it will never improve. It says in the Scriptures, "... every good tree bringeth forth good fruit, but a corrupt tree bringeth forth evil fruit...." You say, "Yes, but the Alchemists of old knew how to transform lead into gold." There you are wrong, the lead was not transformed into gold, it disappeared, and gold took its place! The personality will never be divine, it will vanish completely, leaving the individuality in its place.

In another lecture I explained to you that the physical, astral and mental bodies (which correspond to the personality) will disappear in time, and *in their place* the three higher bodies will manifest their light and beauty. The personality is only a receptacle, no more. Whatever you do to try and improve it, it cannot rise above its egocentric nature; when it is no longer egocentric by nature, it will no longer be the personality!

You must not count on the personality. Use it, put it to work, control it... that will make it take a back seat, and then there will be nothing to stop your divine side from emerging in all its glory and power.

People often think when they are in a negative state that their personality has become worse and, when they are in a good state, that their personality has become better... this is not so. The lower nature never improves, but the higher Self is always there waiting, always ready to manifest if given the right conditions. Then the personality will take over again and get you into a dreadful state, and so on back and forth ad infinitum. You must understand that neither self is subject to change, neither becomes worse or better... no, the self is always the same, but two opposite natures express themselves alternately according to the conditions the person creates for them.

The individuality is never negative, gloomy or selfish; if man manifests those things it is not the individuality which is the cause, but rather the personality. Neither nature goes from one state to another, good does not become bad nor bad become good. Each one remains as it is throughout Eternity.

A person who does something marvellous is able to do so because he has left his personality behind for a moment; upon his return he finds everything as he left it and, as he usually identifies with the personality, he cries, "I am still the same!" No. He is identifying with the personality! If he identified with the individuality, he would see that he can keep on accomplishing wonderful things all the time. His mistake is to return to the level of the lower self after praying, meditating, contemplating, experiencing glorious moments and doing something truly magnificent, and to think, "But I am always the same! I cannot advance, I cannot improve myself." Who did the marvellous things you have been doing, who lived those rapturous moments? Not your lower self! You see how many things are not clear yet in your heads?

Suppose you have been meditating and praying, you have been in the light, far from all ideas of lust, when suddenly a pretty girl walks by and, because the same ideas and images come to you as before, you say to yourself, "How can this be? While I was meditating and praying, I was entirely detached from all that!" Yes, your individuality is always detached from "all that", but as soon as you leave it and go back to the personality, what happens? The personality is still there, very much alive, and it comes to the fore and manifests itself! If you go by a restaurant when you haven't eaten all day, it is only natural if the appetizing smell of food appeals to you!

In times of peace the inhabitants of different countries are smiling and affable, nice and kind to each other, but let war be declared by either country and see how capable they are of destroying each other! Was it the people who changed? No, the two natures manifested through them according to which conditions were favourable at the time. Or take a young girl who is absolutely innocent, pure and virginal, and put her down in certain situations: you will see what she is capable of doing! It may remain somnolent for a long time, but the personality asserts its rights sooner or later.

There, if you have understood what I have told you today, it will be a source of enlightenment. People think they can change evil into good, but they cannot, it is either one or the other. When good is manifesting itself, evil is nowhere to be found, but if the good should weaken, you will soon see that evil is far from dead. Evil is not eternal, that is evident, and it is sure that one day it will indeed be completely changed and transformed, but that is a cosmic undertaking and only Cosmic Intelligence can decide when and how it will be done. In the meanwhile evil fulfills its mission, which is to teach mankind some hard lessons... only, as human beings we cannot discern what the Cosmic forces' intentions are. We imagine that evil will be there forever, and good also, and that both will always be at war with one another; we think evil is as powerful as God and stands up to Him, and that God is so upset and bothered by this that He asks human beings such as knights and crusaders to come and help Him in His battle against the devil! That is what most Christians think.

Evil will not last forever. It exists because God has given it the right to exist, but as soon as God tells it to disappear, it will disappear! Whereas good, and only good, is eternal and everlasting. Evil is temporary, although humans are not able to make it disappear, they are too feeble, too fearful, too ignorant. God alone has enough power, He is the only one who can decide to do away with evil and replace it with good. These are difficult ideas to understand, I know, but by meditating and praying you will attract the light and that will make the Friends in the invisible world want to help you.

No exercise, no method, no yoga, no formula can improve the personality: it is what it is. The crude matter it is made of and its roots buried deep under the ground keep it in constant touch with the lower entities and forces, and they feed it and encourage it and tell it what to do. That is why it is so egocentric, so cruel and coarse, so disloyal. Its roots are immersed in the earth's entrails, where evil is. The individuality's roots are in Heaven, it lives in divine regions; if we learn how to identify with it, little by little it will take possession of us.

The goal of all spiritual exercises is to help the individuality to install itself more and more each day. You can meditate, pray, make sacrifices and try with all your might but you will always be the same basically, that is to say... but I would prefer not to say how we remain the same! In spite of all the magnificent things we do! Yes, the weaknesses are still there, the same habits are there, the same quirks, the same vices and sadness... why? Because the personality is still there. All those wonderfully successful things you have done, you were able to do because you allowed the individuality to manifest itself... but as you continue to live close to the personality and to identify with it, you think you have not changed, and that makes you sad.

You have devoted thirty, forty years to exerting yourself, you have tried everything, and you are dismayed to see you are still the same. You should not be upset! Instead say to yourself, "I have not yet managed to control my personality, but I know why: because I have put it first, above the divine side, the side of me that can make everything better." And if you really decide to change your attitude, you will transform yourself. Yes, but on condition that you put the individuality in its rightful place, for unless you do, it will be there for a minute, a day or two now and then, it will gently murmur a few words, create something beautiful, and then be chased away by the personality. From time to time, under the inspiration of the individuality you achieve something wonderful, extraordinary, but that does not mean you are not still capable of doing harm, of being evil. In fact everything magnificent that man has produced has not been accomplished by him but by something else that manifested through him. All the world's great artistic, poetic, mystic works of art... the poor personality is incapable of accomplishing those things, it can only furnish some of the material. All that divine beauty comes from above, from the higher Self.

When the personality is effaced, when we are deaf to the urgings of the lower nature, another being will come to live in us: we who are in the middle between the two, will die to the personality and live only for the individuality. What is "we", what is "I"? That is a mystery. We are neither the personality nor the individuality, but something other than even those two. You cannot know in the space of one minute who man is, what we are made of... it is not possible, it is a great mystery. But we will talk about it again, and little by little the light will come.

Some of you will say, "The more we study this Teaching, the more perturbed we become, because we have the impression we do not really understand anything." I know, that is the way it is for a while, but as you will see, it is only a passage. Many who come here realize that they were much happier, much more satisfied where they were! But once they are imbued with the Teaching and become enthusiastic and inspired, they decide to transform their lives and lead a divine life. At that point nothing works, you can see them fermenting. I say to them, "Be patient, you must go through this fermentation, as in Alchemy." In Alchemy the very first phenomenon is fermentation: the material clouds over and darkens, then it ferments and dies... and resuscitates! That is what happens to those who come into the Teaching. Perhaps not to everyone, for some refuse to change, to make a decision, to choose, and therefore they remain the same. For them everything works well, without any fermentation, but that is not necessarily a good sign. When I see someone going through the fermenting stage I am delighted, I think, "There is someone who is on his way to discover the Philosopher's Stone... he will soon be able to transmute his metal into gold!" So you see it is a good sign if things do not work out for us when we first enter the Teaching!

Perhaps you will understand better if I take an example from the human system. Suppose that because of leading a disordered life you have accumulated a lot of toxic poisons in your system, but you go right on eating and drinking and working, carrying this illness, this impending death around with you perhaps for many years. It does not declare itself for it prefers the dark; it knows that once it is discovered it will be chased away by pills, all kinds of medicine and doctors bent on starving it to death. Therefore it goes on undermining you silently without being seen. But when you enter the Teaching and begin to observe certain rules and practise certain purifying exercises, you come down with fever, colic, headaches and everything else, because the system, or rather the good entities that live in it, take their courage in hand and say, "The time has come to chase away these blokes, these evil creatures who have settled here and live at my expense!" So the system shakes itself and makes a tremendous effort to get rid of them. Those who are watching without knowing what is happening think it is the Teaching that drives people mad! No, the Teaching is the cause of the desire to be free of the undesirables installed in your system.

The same thing happens when you make up your mind to fast for a few days in order to be purified... palpitations, dizziness, nausea, migraine and other symptoms attack you when you fast, for at that time the system becomes a diagnostician and indicates where there is trouble, where your weak points are. Those who don't know this are frightened, they stop at once and say, "Ah, that's the last time I'm going to fast... it kills you!" A fast is just what you need to rid yourself of all the waste matter your system has accumulated.

I ask someone who is in prison, "Why are you here?" "Because society is all wrong, because humans are cruel and mean, because I was betrayed by so and so...." I say, "No. The reason is that you have too much faith." "Too much faith? In what?" "In yourself", I answer. "In your way of looking at things, in your convictions, your calculations. You were sure of succeeding, of being right, of escaping justice, of never having to pay because you believed implicitly in yourself, you relied too much and hoped too much and believed too much in yourself.... If you had believed less blindly you would not be here at the moment!" Yes, those who believe too much in the machinations of the personality finish badly, for as the personality is lacking in foresight and nearly blind, it will eventually lead them over the precipice. I say, "Do not hope, do not believe in yourself, *doubt* in order to be saved!" That may seem odd to you as far as moral advice goes, but everything, yes everything, is going to be turned around and set right, believe me! I am a destroyer, a demolisher, Master Peter Deunov told me so. Don't misunderstand, I do not destroy everything... never do I destroy anything divine. But when I see something old and worn and mouldy and moth-eaten, I take great pleasure in destroying it, believe me! If I can succeed in removing a tumour, a bit of gangrene, the way a surgeon cuts into the tissues and removes only what is ruining the system, I am drunk with joy! I present myself: the world's greatest demolition expert!

Now, if any young person comes and tells me they feel like demolishing the whole world, I say, "Is that all, child? That's nothing! Come and learn from me, I will tell you how to use your hammer so as to do a good job! What you are doing is nothing at all!" And I will put them to work doing a fantastic job of demolition!

I wonder if I have succeeded in persuading you. The personality is so tough. You can fry it, boil it, cook it as long as you like, it is always there! From the bottom of the pot it will look up at you and say, "Here I am!" It is really extraordinary.

Le Bonfin, 17 August, 1971

Chapter 7

Which Life?

The Master reads the Thought for the day:
"We all have a higher Self in Heaven. In the silence we come in contact with the strength, power, harmony, light and abundance of this soul... which is the Quintessence of God Himself."

I have often talked to you about the higher Soul or Self called the Buddhic plane or body, and the lower soul or self called the astral body. Both are the seat of emotions and feelings but entirely different in quality, nature and power, for the higher Soul is an aspect of the individuality and the lower soul is an aspect of the personality; that is the important thing to know.

I was talking yesterday with a group of brothers and sisters who had expressed their amazement at the elucidation, the tremendous amount of light I had just projected in my lecture on the subject of the personality and the individuality. That's nothing yet! You may think you have a clear idea of the two natures of man but if you knew how much more there is to reveal! What do we see when we look at the majority of men, when we observe their life, their way of working and their ideal, the goal of their existence? Everyone does their best to fy their desires and ambitions, without ever wondering about the quality of these desires and ambitions. Do you know many people who turn to God and say, What is Thy Will for me, O Lord? Am I in accord with Thy plans and projects? Am I too concerned with my own will rather than Thine? What is Thy opinion, where and how can I best serve Thee? Not many.

The personality urges man to live his *own* life, to plan his life around his own desires and wishes. That is the personality and the whole world works to satisfy it without ever asking whether there might not be other goals to realize a thousand times more important, more glorious, more divine. There are exceptions of course, but they are treated like fools or cranks, not to be taken seriously. The individuality, on the other hand, wants above all to know what the projects of Heaven are and then to realize them. A man's life is changed once he thinks this way, rather than how he can satisfy his appetites and lower leanings whilst continuing to live in the midst of illusion. When he starts to follow God's plan for him, when he starts out on the Path, then he begins to live the real life. It is hard to know what God's plans are, but we could try asking Him, we could implore Him, "O Lord, even if I cannot understand, even if I have no way of knowing what plans Thou hast for me, still guide me, put me on the right path, make me do Thy Will even unawares, even blindly. Use me as Thou wilt, make myself Thy dwelling-place!" We cannot always tell what God's Will is for us, but we know the general direction to go in: always toward the good, toward love, abnegation, uninvolvement, impartiality, kindness, generosity, sacrifice... but as we humans lack clairvoyance, insight, lucidity, we should put ourselves in His hands and say, "Lord God, Thy Will be done despite me, send me where Thou wilt!" In that way we realize God's plans blindly, unaware of doing so. We think, "How did I save all those people when all the time I thought it was the wrong thing to do! It turned out to be the

greatest good! What force took over, what power used m
It is not always given to us to know when we are being useful.

You must all supplicate Heaven to accept your services.
Say, "Here I am, I understand at last, I see that I can do nothing with my lower nature. It remains too headstrong, too
tough to change. Yes, after wasting all these years, I finally
understand, Heavenly Beings, that there is nothing to be done
with it. Our Master told us not to expect anything from it, and
now I see it is too limited, too blind, too ill-intentioned. I ask
you to send me in exchange the most wonderful, perfect creatures to take its place, to install themselves in me and guide
me, teach me and take control of my life so that in spite of
myself I may realize God's plans."

That is the best prayer in the world! All others contain
some personal interest, some selfish motive, you want to win
God, coax Him for your own purposes. But in that prayer you
put your cards on the table and say, "Here I am, Lord, I am
prepared to die, take my life, erase me entirely if you will, but
send me Heavenly Beings to replace my lower nature." Then,
fair exchange, you have paid out what is your most precious
possession and Heaven is obliged to hear your prayer and
grant it. Yes, even Heaven must pay in order to obtain something. Nothing is free. If you think you can obtain Heaven's
blessings whilst continuing to amuse yourself, behaving irra-
tionally, selfishly, you are mistaken. There are quantities of
people in the world who pray to obtain material advantages
and so on, and of course these prayers are not granted; but if
you give your whole soul and beg for wisdom, love and peace
in return, those above are ready to give you everything! As in
a pawn shop: you give them your watch, your ring or whatever, and they give you a few coins in exchange. Everything,
dear brothers and sisters, everything is a reflection: As above
so below.

Today I have given you the best prayer. There is nothing
to be done about the personality, nothing. Even though it

ᴊ learns a lot of things, becoming very clever and ef-
ᴊ, even going so far as to create beautiful things... it will
ᴊ be what it is, that is, thinking always about its own profit.
ᴊt may throw gold dust in your eyes with its importance, but
its selfish nature works always for itself, never, ever, for Heav-
en. It will not change. If it did, it would no longer be the personality, it would be the individuality! That is the goal of Initiation. In the Initiations of the past, the ritual was for the
Master to enclose the disciple in a sarcophagus for three days
and three nights in the trial of death. Death was of course
symbolic: the lower nature died in order for the higher nature
to come to life and for the new man to emerge, entirely under
the control of the individuality.

From now on, why not decide to speak to the Heavenly
Beings, to ask them, "Have you plans for me, am I going
against these plans? Perhaps I have done nothing since I was
born but upset God's plans!", and then ask for forgiveness,
call on the Light, on the Heavenly Beings to come and install
themselves and direct your life. If you are so entangled, so
caught in the claws of the personality that you cannot understand what I have just said, then there is nothing to be done,
you will have to be unhappy, you will suffer, and you will rip-
en! The personality is a pumpkin, it will never be a water-
melon, but as a pumpkin, it will ripen. You say, "Well,
plants can be grafted...." Yes, but grafting is something else.

To graft the sweet fruit tree of the individuality onto the
wild crab-apple tree of the personality, that is the work of the
Initiate. The personality furnishes the sap, the invigorating
quintessence which the individuality transforms and uses. In
the beginning the disciple is a wild pear tree or a wild quince
with bitter uneatable fruit, but then the Master comes along,
grafts on a branch of individuality to transform it, and it becomes a splendid tree! The disciple supplies half of the material and the other half grows, flourishes and bears delicious
fruit.

The light which I have thrown on this subject of the personality and the individuality today will make it possible for you to tell exactly what and where you are. You will know the truth. Others may say to you, "You are this or that", but they can be mistaken, whereas you will know where you are according to the standards of the Teaching. If you do not apply these rules, the divine laws, and think only about satisfying your lower nature, make no mistake: the crowd's applause and praise because of your success in one field does not make you a divinity! Far from it. Let the crowd think what it wants... in any case it is blind, it doesn't know the true value of things. Do not go by the crowd, do not trust the crowd. Whether they bear you in triumph on their shoulders or throw tomatoes at you is not the criterion; you must have your own absolute, unchanging criterion.

If you work for God, for all Heaven, for the Kingdom of God, whatever happens to you, whatever people may say about you, however they may treat you, you will never be discouraged, never afraid, you are on the right path, that is sure and certain.... If you give up it proves that your goal was not a divine one but selfish and personal, that you will neither risk anything for the truth nor work for truth... only for yourself. All those who have worked for the truth, for the realization of God's plan, were never afraid, no matter how persecuted they were, no matter if they were put to death they scorned all that because they knew they were immortal, they knew that glory awaited them.

Without this light and this force of character one is always hiding, always camouflaged. Why? To gain a few little odds and ends; a little satisfaction for one's vanity, a few favours from the crowd, passing favours.... Ah, it is extraordinary how people do not know how to analyse themselves! Many people are afraid without realizing it, they have never heard the saying that the fearful will never enter the Kingdom of God! What is the Kingdom of God? You can think of it as a place above inhabited by the elect, the Angels, or you can think of it as an inner state of peace, of harmony, of light... it is always the Kingdom of God, both outside and inside us. Cowards will not be admitted to either.

The terrible thing about this Teaching is that even if you want to fool yourself, you cannot. Terrible, isn't it? That is why, if you want to keep your illusions about yourself it is better not to come to the Fraternité because you will never stop saying, "Why did I do such a stupid thing again? Why can't I control my thoughts and feelings?" Without the Teaching we are proud and puffed up, we think we are the crown of all Creation, but the more the Light grows, the more we are obliged to humble ourselves and the humbler we become the more reasonable we are. "Look where I still am! Ah, please Lord, take pity on me and help me."

Now, you should know that the centre of the personality is the astral world, that is where the suggestions and impulses come from that influence you in a negative way. The astral plane gives the zest and the mental plane follows with arrangements to gratify the personality's slightest wish. That is what you must understand. Our behaviour, our conduct, is dictated by the astral world. The mind, the intellect, is superior to the personality and perfectly capable of stopping it and imposing its will on it, but instead it waits on it hand and foot. Is that not true? Look, everyone uses his mind to help satisfy his desires and passions and lust. All our schooling and instruction, all our learning, all our cultural wealth is in the service of obscure, bizarre, sombre ideas that come from who knows what underground cavern.... The most learned and informed people, supposedly the most scientifically enlightened, are working for forces and impulses that are far from clear. That is the sad truth, if you don't believe me go and see for yourself!

When the astral body starts working for the mind, or better yet, when the mind works for the spirit and the soul, then it will be perfection. And that is the role of prayer: to submit the physical, astral, mental bodies, that is, the lower triad which thinks, feels and acts selfishly, to the higher triad which thinks feels and acts also, but divinely, for the whole world. To ask the individuality to take over one's entire being, that is the best prayer. As long as the personality is there trying to impose itself, playing the boss, the big expert, even if the individuality is able to slip some good advice in from time to time, or give its blessings, it cannot remain, for the personality holds the power. That is why nothing works out. Yes, the individuality manages to help us now and then by projecting divine sparks on us, inspirations that dazzle us, but it does not last, a moment later it has retreated, because the personality is still in first place. This mysterious unknown quantity, the human being, is in between the two natures, it is therefore up to him to pray and beg the individuality to take the place of the personality once and for all!

This question has never been treated this way; I present it to you today because the time has come for you to say, "Here, you... hurry up and let yourself be replaced by the individuality, what are you waiting for? You still hope, you still believe in the lower self? But that makes no sense! Now, decide!" And you will see, Heaven is neither deaf, nor cruel, it will send guides and consellors to help you get away from the claws of the personality.

Le Bonfin, 19 August, 1971

Chapter 8

The Image of the Tree
The Individuality Must Consume
the Personality

Today I will give you an image that will throw a little
more light on what I have explained already about the personality and the individuality: the image of the tree.

When you study the structure of a tree, you see that by
analogy the roots, trunk and branches correspond to the personality, whereas the leaves, blossoms and fruit correspond to
the individuality. The personality represents the earth, matter, and plays the role of the container or conductor, whereas
the individuality plays the role of life, the spiritual element or
Spirit which is behind all manifestation. Both are indispensable.

As the tree grows, its roots dig down deeply into the
ground, the trunk expands and thickens, the branches spread,
it becomes a large and powerful tree, towering up toward the
sky. You have all seen forests full of huge and magnificent
trees with roots that stretch far and wide and branches that
climb to extraordinary heights... like the sequoias I saw in
California... I have never seen such trees! It was splendid,
they were so thick and tall it was unbelievable, and they were
4000 years old! My heart rejoiced to see them.

So the tree grows, spreads, and becomes strong and beautiful while the roots, trunk and branches carry out their func-

tion which is to support the leaves, flowers and fruit. The roots, trunk and branches are there all the time, during all seasons, whereas the leaves, flowers and fruit make their appearance and then drop off. In the same way the personality, made of our physical, astral and mental bodies, is always there as a permanent support, whereas the individuality or spiritual side, the joy and inspiration formed by our Causal, Buddhic and Atmic bodies, come and go.

Take the vine, black and bare in winter but beautiful and thick with leaves and grapes in summer... that is a human being, you can help him to grow strong by feeding him, you can make him clever and learned, but no matter what you do the personality will always be there. As soon as the individuality or Spirit is allowed to manifest itself, then the vine blossoms and gives fruit. You can also take as an example a hose, or an electric wire : their function is to conduct water and electricity, but by no means will either the hose become water or the wire become electricity... they will always remain conductors.

By taking care of it you can render the personality extremely capable, powerful to the point of standing in the way of the sun... but it will never be the sun. You say, "That's not what we were taught... we were told that with a lot of effort and work our lower nature can be made divine...." Never. All it can do is disappear and leave room for the individuality. One day the physical, astral and mental bodies will disappear, and their place will be taken by the Causal, Buddhic and Atmic bodies... then man will be transformed into a divinity, the earth will lose its density and weight and become transparent and radiant, and all will return into the bosom of the Eternal. That is the Initiatic Teaching. Oh, don't worry, it won't be for a million years or so ; it will take so long for it to happen that you will wish it were faster!

Now, there is another image in the field of alchemy. The nucleus of an atom made of lead contains 82 electrons and 82 protons, whereas a gold atom contains 79 only. To transform lead into gold therefore, you have only to subtract three electrons and three protons... extraordinary! Yes, take three electrons and three protons from the lead atom and the gold that is hidden behind will reveal itself, as a mask that is removed discloses the real face. Take away some of the lead's covering, a few particles of the lead's atoms, and you will see that the lead is actually gold!

It becomes gold because that is what it was made of originally. This is what happens to human beings: remove the personality, the three particles of the physical, astral and mental bodies which prevent the individuality from manifesting itself, and you will discover perfection: man is made of gold. If scientists could carry out this operation on a grand scale, they would produce mountains of gold from lead! The process is too costly at the moment, but perhaps later on there will be other ways of doing it, it is not impossible. What interests me is the interpretation of these symbols, lead representing the personality and gold representing the individuality.

I would like to insist once again on this point: instead of fighting eternally against the personality in order to educate it (which is impossible), it would be better to busy yourself with the individuality, to attract it and give it every chance to install itself in you. In that way, one day it will be so powerful there will be not a trace left of the personality, and you will be able to say, as did St. Paul, "Not I that liveth, but Christ who liveth in me." It is no longer you that acts, that is, your lower self, but Christ who acts through you.

Continue to manifest the divine qualities of unselfishness and love and, one day, you will see that your personality has been swallowed up and digested... I use those terms because I know a law that you may not know, which is that every creature absorbs some other creature, that is how he survives. Plants on the ground eat minerals, animals eat plants, man eats animals or at least their produce... and who eats man? For there are entities who do eat humans or rather their emanations. This law applies to all beings on the ladder, all the way up to God Himself... who feeds on Heavenly Beings. Yes, God eats Heavenly Beings, He swallows and absorbs them but in such a sublime way and on such a pure level that there are none of the terrors and suffering that eating provokes on the lower levels.

Suffering begins at the animal level. Earth, crystals, rocks, plants, do not suffer when eaten because they have not yet developed an astral body, the seat of sensitivity and emotion. Beginning with the animal kingdom creatures suffer when they are eaten by another creature, but for those who can lift themselves up to heavenly regions, pain disappears and is transformed into joy to such an extent that to be eaten by sublime entities brings happiness and joy, a felicity beyond description. All religions say that man should offer himself to God in holocaust, in sacrifice, so that God can nourish Himself... showing that man's spiritual ideal should be to be absorbed by the Lord in order to live in Him.

If most humans are insensitive enough to enjoy eating the flesh of animals, there are more and more vegetarians who take only their produce such as milk, eggs and no more. In the same way the Higher Beings eat only the emanations of human beings, their best thoughts, feelings and actions, the creature himself remains intact. For Heavenly Beings, it is a most delectable feast.

True spiritualists, having understood that it is highly desirable to be absorbed and devoured by the Lord, offer themselves to Him in sacrifice... but how many really understand what sacrifice is? The word frightens people because it is associated with pain and death. In the lower regions it is true that if man is eaten by the lower entities he is lost... but if he offers himself to the celestial entities he will be enriched, embellished, improved and aided in every way... resuscitated!

We must not be afraid to offer ourselves in sacrifice to Heaven, for in that way the personality is consumed, no trace of it will be left, and then you can say, "Not I (the personality) but Christ (the individuality) manifests in me and directs me. Before, I guided myself, but now I let myself be guided by Christ within me."

With my two ideas of the personality and the individuality, I do no more really than transpose the original Bible truths into another form; I say the same thing, but in a different way. St. Paul was not the only one to be transformed by the presence of Christ, the apostles received the Holy Spirit at Pentecost, and they too were transformed... it was no longer the crafty personality with all its folly, passion and stupidity that expressed itself, but the individuality. For that is what the individuality is, the Holy Spirit, the Higher Self. The Holy Spirit is a cosmic Principle, and the Higher Self which each human being bears within him is a particle of the cosmic, universal element which is One with the Holy Spirit.

The Holy Spirit is within us, but also, in the form of a spark, a quintessence, is the Christ, and so is the Father. The Father, the Son and the Holy Spirit are the great Cosmic Principles, and the Higher Self of each human being is made of the same substance, and is part of the Father, Son and Holy Spirit, with the same divine attributes. We all are a divine spark of the Holy Trinity which is in us but which has not yet taken possession of our being to the extent of manifesting itself.

We will come back to this question in the following talks. For today, it is enough to bear in mind the image of the tree with its roots, its trunk and its branches.

In spring, a tree blossoms forth with leaves, flowers and fruit: the individuality. Why are the leaves, blossom and fruit not always there? They come and they go (like poetic inspirations that visit us for a fleeting instant from time to time and then leave again), whereas the roots, trunk and branches remain all the time. That is why I said you cannot *change* the personality; it can grow longer, thicker, denser, but in essence it is always there, the same as ever, with its roots (the sex and stomach), its trunk (the lungs and thorax) and its branches (the brain). In order to change we have to let the Spirit or individuality pass through us, then, yes, there will be an extraordinary change. In the same way as a tree which covers itself with leaves, blossoms and fruit to everyone's great joy, so a man who lets the current of the individuality pass through him is a source of joy and blessings to all who come near. Like the tree the man grows and develops, but unless he lets the pure, spiritual force and energy course through him, his personality will remain bare and stark as a tree in winter.

Now let us see how the tree corresponds with the different bodies of man. The roots correspond to the physical body, the trunk to the astral body and the branches to the mental body, and these three bodies, physical, astral and mental, form the lower triad or personality, which permits man to act, to feel and to think in the lower regions. Then we see that the Causal body corresponds to the leaves, the Buddhic body to the blossoms, and the Atmic body to the fruit of the tree, forming the higher triad of the individuality, thanks to which man can think, feel and act in the higher regions. The personality and individuality are two triads, and when the individuality is able to enter and dominate the disciple's personality, he becomes the seal of Solomon, a complete being. For two thousand years Christians have repeated, "I am the vine, ye are the branches", without realizing what a tremendous symbol it is. We are a triad, but another triad must come, a divine triad, a Holy Trinity within us, to manifest through the lower triad and make it bear fruit.

The personality represents, as it were, the foundation of the human being, and it is therefore out of the question to destroy it, the man would die, he would no longer exist on the physical plane. It is thanks to the personality that man manifests himself physically and for that reason it must not be suppressed, on the contrary it must be maintained, fed and cared for, but without satisfying its demands, without giving it its head, and above all without forgetting to bring the individuality down into it ceaselessly. Is that clear now?

If you think this subject obtuse and difficult and you refuse to think about it, of course you will go on living, nothing will keep you from going about your business...but you will not evolve. Whereas with the knowledge you learn from Initiatic Science, although you may not hold a prestigious position in society, you will be treading a glorious path. You must understand that these are two different planes. You have graduated from university? Good! You have a diploma, a degree, a good situation, a lot of money and prestige... as far as the world goes you are all set. But what does Heaven think? In any case, it is not scholarly learning that will help you enter the heavenly gates, to open the door to Heaven you need Initiatic, esoteric, mystical knowledge. Perhaps you will then also be given the earth, but it will not be because of that knowledge.

The culture, science and philosophy you gain from mankind do nothing to transform you, they merely allow you to live comfortably in the material world. But the knowledge you receive from the Universal White Brotherhood transforms you, you are no longer the same. Is this clear now? You must choose Heaven or earth. If I am asked to choose, I will answer, "Both, please!" You know the story? A shepherd asks a starving beggar in rags which he prefers, bread or milk, and the beggar replies, "Both, your Lordship!" Some people choose only the world and others choose only Heaven, but that is no good, without the earth Heaven vanishes, without Heaven the world abandons you... and you lose both! Therefore if the Lord asks me, "Which do you wish?", I will answer, "Both, please, Lord."

And now, good afternoon, dear brothers and sisters. Are you happy? I think so, by your faces. And yet I have given you neither money, nor castles, nor swimming pools, nor cars, nor women... and still you are happy! How can this be? Very mysterious indeed.

Le Bonfin, 28 August, 1971

Chapter 9

Working on the Personality

In a number of preceding talks, I have said that you must never accept the personality either as an advisor or as mistress of the house. This first method I gave you was to help to overcome the personality. In the beginning it is the best we can do, while waiting to reach the higher levels of comprehension. But I warned you that with this method you would never have the last word because the personality is too clever, too guileful to give up, it will go on plotting its triumphant return as mistress of your house. As soon as you relax a little and show a little less clarity of mind and vigilance, in it comes, ready to go to work again. And it succeeds! Once again you are at its mercy.

But it is not a bad method actually, because it forces you to analyze and struggle with yourself, thereby developing discernment and willpower. But it is never totally effective because the personality even when it is your servant is not faithful, it will always be crafty and artful, rebellious and malicious. Only in appearance does it accept to be your servant, whilst actually trying night and day to reverse the situation and take over the power. Here is an example : two countries go to war against each other. In the end one of them wins and the conquered country must submit to having its land occupied, to relinquishing its freedom and paying heavy taxes to the enemy. Yes, but the situation is not irrevocable, it will not last forever; the conquered people will not go on accepting defeat and will go underground to work to liberate their country. Nothing official of course, but everything the enemy does is undermined until one day when he is lulled into thinking himself in complete security, sitting back on his laurels and proud of himself... that is when the surprise explodes, and again the situation is reversed. These phenomena have occurred in the economic and political history of all the world's nations, and on all levels of life in one form or another.

This is what happens in the inner life as well. We will never entirely subjugate the personality, it will remain alive through thick and thin and go on working underground... it will never give in. To keep it in a state of submission you have to be armed to the teeth, vigilant at all times, which is exhausting. Even saints and Initiates tire sometimes and relax their hold... and the personality is there to bite them! It is extremely tenacious: you uproot it and right away it grows back in. You know what couch-grass is, don't you? Well, the personality is like couch-grass. For that reason I will now give you another more effective method.

Do you remember I used to tell you that we are not really very well equipped to fight against evil, whereas evil is extremely well-armed, it has a huge arsenal behind it and before this force we are powerless. For such an uneven battle there is only one thing to do: ask God to become our ally. What do countries do when faced with war? They form allies with other countries. It is instinctive with man to turn to others for help when he knows he is vulnerable. We too should realize that evil has so many resources behind it, so much knowledge and power, that we will never be able to conquer it alone. The solution is to form a link with the Lord, with Heavenly Beings, Archangels and Angels, and ask them to fight for us.

Then we can be spectators observing the battle, whilst Heaven carries off the victory! Heaven, that is, the divine side in each one of us, is all-powerful. But who are we to stand up to the power of evil?

When I was young I did not understand this. I had been taught, I too, to fight against evil, to try to extirpate it, annihilate it, and I was torn apart by my efforts, literally dislocated, because it is too much for a man to fight alone against the terrible, mysterious inner forces of evil. Later, when I began to think about it, I realized that there were other, better ways of doing it. That is why I now say to you, stop struggling to overcome the evil within you because it is not possible. You may win but you will kill yourself in the process... could you remove your intestines and remain alive? We must use a different method.

Do not try therefore to eradicate your personality, not only will you be unable to, but it is you who will be eradicated! First form the link with Heaven and then handle the personality with the absolute conviction that you are really its master: it will be forced to obey. Remember to do it boldly, with conviction, otherwise it will be the one to overcome you. Take for instance, a rider who is afraid of horses: the horse will throw him to the ground... the horse feels his cowardice and wants to teach him a lesson! Alexander the Great had a great horse called Bucephalus who was indomitable, no one could mount it but Alexander, the horse sensed who Alexander was. The personality is a wild horse. That is what the Initiates meant by "to dare": dare what? Dare to dominate the personality. People dare to steal, to cheat, to commit murder, yes, there they are courageous, but when it is a question of overcoming the personality, they are afraid.

And even when you have succeeded in taming your personality and it has been replaced by the individuality, this does not mean it will disappear completely, it will always have its roots in the physical body, because the physical body is the last refuge of the personality. Even when the astral body has no more selfish desires, even when there are no more dark thoughts in the mental body, the personality still has its foundations in the physical world, it continues to survive. This is necessary because if it disappeared from the physical plane, neither the individuality nor God Himself would be able to manifest. This replacement, this substitution is therefore neither definitive, nor complete. On the psychic plane, the personality can be replaced once and for all, but it will still remain on the physical plane. You can compare this to a change of personnel in a bank, a university, in any administration... the building remains and everything else functions as usual, but some of the people have gone and others have taken their place.

In order for you to understand what I am saying a little better, I will repeat what I said one day on the subject of memory. As you know, the cells of the body are constantly being renewed, and at the end of seven years, it is said, new particles have entirely replaced the old ones. Yet the man is no different, he continues to do stupid things, he has the same faults, vices and weaknesses, he comes down with the same illnesses. This would seem utterly inexplicable if we had not discovered the fact that although the system takes new material from the exterior to replace the old, worn material, one thing remains that is not renewed: the cell's memory. Each new cell goes on working exactly as the old one did.

It is the same thing in offices, in factories and so on. When old employees leave they are replaced by newly-hired, younger help who are given exactly the *same work* to do. That is what I mean by memory. The new ones inherit the same work and carry on with the same methods, in the same tradition, they have the same goals, the same objectives... the work goes on as before! In the human being it is this memory of the cells that makes him go on making the same mistakes and doing the same stupid things, because, although he has replaced the particles of his body he has forgotten to change their memory, and that is what goes on being transmitted from generation to generation. It may not be the same cells, but they do the same thing, they react the same way and make the same gestures, for the same purpose.

Now when you pray, "Almighty God, I beg you with all my heart to take over the direction of my life in place of my personality", you act not only on the material particles of your physical body, but on the memory of the cells, their ingrained habits... which are then replaced with new and better ones.

What I am telling you here is most important. Even if you manage to control your personality, even if you force it to carry out your idealistic plans and projects and make it do as it is told, it will not be the same as if you had removed the old "clichés" and put new "clichés" into its "head". It may bow to your will, but that doesn't mean it has adopted your imprints, on the contrary, it still has its own, and is waiting for the moment to do what it wants. In beseeching God to take possession of you, you are robbing the personality of its old memory and habits and from then on it will no longer be the personality that guides your life but the individuality! The form is still the physical body, with a stomach, lungs, brain, etc... but with new contents.

St Paul said, "I am crucified with Christ, nevertheless I live; yet not I, but Christ liveth in me...." Christ manifested through the man called Paul, that is, through the part of his personality that had not been made to disappear. When the personality is replaced, only the contents are different; as with stuffed animals, the entrails are removed but the animal still has the form of a lion, an eagle, a bear or a sparrow. When the individuality installs itself there will be no change in you, you will still be the same fellow everyone knows, but your *memory* will be replaced. In your heart of hearts you will be so changed that everyone will notice the difference in your emanations... that is the marvellous thing, you are the same and yet you are completely different!

Do you remember the Gospel story of the Transfiguration, when the face of Jesus became so luminous as to be incandescent? This phenomenon was the result of the contents, the spiritual Guest who visited him at that moment and manifested through him, but his form (his features, size, appearance) was unchanged. The form does not disappear. When Jesus appeared to his disciples after his death in his Body of the Glory, the body of Christ, he appeared in his usual form so that he would be recognized. Besides, when you die and go to the other side, you keep the same physical appearance you had on earth.

The Cosmic archives conserve the bodily form people had during their lives, sometimes even their clothing. Suppose you are at a spirit séance and you wish to materialize someone who lived several thousand years ago... if he comes and talks to you he will have exactly the same form he had in the past. Actually it is not he who comes, but his form which has been in the archives of Nature, an animated form which speaks to you and which you can touch, etc.... Forms do not disappear. The human being evolves, advances, and discards one form in order to take another, but each one of his forms is carefully preserved in the world's archives, the Akasha Chronica. A form does not last beyond one incarnation, in his next incarnation he will have another form and that too will be preserved. All forms are kept for millions of years, perhaps until the universe disappears, and always there are new ones. Does that surprise you? It is so, nevertheless. Even if what I am telling you seems strange, do not worry about it and simply believe me, for I am not inventing these things. You are victims, the product of a materialistic education, in a materialistic world which has led you into error, but in time you will correct your vision of the world and in the end see things as they really are.

And so, dear brothers and sisters, I might as well tell you now that you will always be hearing this same tiresome subject when you come here... the personality and the individuality... for this knowledge will make it possible for you to transform your life. I know you would prefer to hear other things, such as the arcane Mysteries of the Kabbala, or the Science of Magic. Well, no, you are faced with someone who has come for the express purpose of being a pain in the neck so that you will have to go to work on your character and change your way of life! If it is displeasing to you, try not to mind too much... a little bit, but not too much! For I have been given this Mission, this Teaching, for you, and I must carry it out to the end. Of course I could talk about all sorts of subjects, there are plenty of other things to talk about, but would that change your life? I wonder. Whereas if you change the way you *live* by applying the laws I give you, all the rest will come naturally, all learning and all knowledge.

Otherwise, this is what will happen: you will read all your books, you will record their contents, but it will only last a while, and in a few years it will all be gone. Why? Because you will have lived in such a way as to dissipate your knowledge. It is such a waste of time, so useless to acquire knowledge that will be forgotten so quickly! If you *live* the right way... I mean, if *we* live the right way (I should correct myself, should I not?), if we live as we should, then our memory will begin to awaken and bring us all that we learned during thousands of other incarnations, and we will remember. We remember all kinds of things, with no need to read or study, we remember: that is the real memory. Note it down and never forget it. For the person who has learned how to live divinely, all the Ancient Wisdom he has accumulated and recorded in his memory comes to the surface when he needs it. He possesses Cosmic Wisdom.

That is what I count on, uniquely: the way I live. If I can manage.... Ah, Lord God! If I can succeed in living harmoniously, in unison with all the spirits of the living Nature, I know I will then have Cosmic Wisdom. It will come, I will recall it, I am certain of it.

Le Bonfin, 30 August, 1971

Chapter 10

The Personality Keeps
You from Reflecting the Sun

Reading of the Thought for the day:

"When we are here all singing together in perfect harmony, as you may have noticed, one can feel the presence of divine entities who are attracted by this harmony and go among us distributing flowers and gifts. Make every effort towards harmony, use all your will to bring it about, so that Heaven will come, for when It does, you will witness extraordinary things, you will feel a tremendous happiness that you won't be able to keep to yourself, and such powerful currents will course through you that you will tremble in ecstasy!"

Yes, dear brothers and sisters, you must take these words seriously, for not only is it an extremely beautiful thought but it is the truth. It depends on you, not me, whether Angels come among us or not. Whatever I do is for me; for it to be your achievement, you must make the effort to enlighten and enlarge your consciousness. Of course, for those who are deeply mired in prosaic and material preoccupations, Heaven is likely to remain many light-years away, but It comes close to those who consecrate themselves to serving it. Night and day It is close by! That is what we must understand: Heaven is not really so far away, it is we who put obstacles between us, and the heavy layers we have formed do not allow us to enter into communication with It. Actually the happiness and joy of Heaven is all around us. If you purify yourselves and make your bodies subtle and receptive and sensitive enough, immediately you will see that what I say is true.

The divine world is not far off, rather it is nearby, and yet much can separate us from it, as though it did not exist. There are people who, no matter what you do or what you reveal to them, do not feel that God exists, nor can they accept the fact that Heaven is filled with luminous shining Beings. They don't feel the order, the harmony, the love; they say, "No, I don't feel anything, I do not believe any of that." What on earth did they do in their past lives to be so insensitive, so uncomprehending? There are people who do understand, at least intellectually, that Higher Beings exist, and that Cosmic Intelligence has created higher Laws and a higher Justice. Others do not accept this as fact but at least they sense these things, they have an inkling of certain experiences. On a higher level are those who both feel and understand. And, in a category by themselves, are those who, because they are very highly evolved, not only understand and feel these truths but act upon them in the physical world, thus making the Truth accessible to others all over the world.

To enter into communication with God, the personality must be kept well under control, for it is the personality that keeps us from understanding and experiencing the other world with all its infinite joys and beauty. The personality is so strong, so overconfident and dominating in most people that it obstructs even the rays of the sun! Like the giant bird Roc in *The Thousand and One Nights,* which became so tremendous when it spread its wings that the sun was eclipsed! All fairy tales are somewhat exaggerated but it is true of the personality that it is so overpowering, so possessive, so all-encompassing that it obstructs the sun. It takes itself for the sun! It wants to be the centre of the universe and cannot stand anyone who does not revolve around it; it has an exaggerated idea of its importance, its rights, and everyone must bow down and wait on it hand and foot. When this is not the case, it becomes bitter and resentful, and seeks revenge on the whole world.

Another fault of the personality is that it cannot foresee what will happen. It expects to triumph by pulling the bed covers over to its side, when it is actually the opposite that will happen. Only at the cost of much useless suffering and expense does one finally realize this idiosyncracy of the personality... when it is too late. The personality is nearly always mistaken in its calculations and previsions, it can never see the future, never, it is blind. You are in for trouble if you go along with the personality... I have seen it so often!

I advise you to think about the personality and the individuality because to me the question is absolutely essential. Many great thinkers have thrown a lot of light on all kinds of subjects, whereas I have devoted myself to shedding light and clarity on only one subject: the personality and the individuality, the lower animal nature and the Higher Nature. The instinctive nature, if you prefer, and the divine nature. That question has been my life's work because I believe it to be essential, yes, it is the key to solving the problems of mankind. If all men knew about the personality and the individuality, mankind would make astounding progress... but people do not care about these things, they prefer to study all kinds of useless things. That is why nearly everyone follows the personality, and breaks his head in the process. The individuality manifests through them occasionally of course, but they do not know when or why, as if in spite of them, in spite of their ill will, Heaven slipped in from time to time with a few inspiring thoughts... but so rarely! If they were conscious, if people accepted voluntarily to pay more and more attention to the individuality, they would progress with extraordinary rapidity along the path toward Truth and Beauty.

Unfortunately people do not calculate correctly. They think if they work with the individuality they will lose all sorts of advantages and gain nothing in exchange, and so they are afraid. It is because of this fear and ignorance that the personality can stride ahead and prevent them from seeing clearly where their interest lies. It would be in our interest, for instance, to forget ourselves for a minute and go to work for the good of all mankind, the entire universe! Perhaps at first this idea will have no appeal, but if you go to work in that direction, you will feel a force, a power pushing back the horizon for you, which will bring you joy beyond description.

Now, when I say we must work for the good of mankind, it does not mean we must love everyone with a total lack of discernment and form relationships and partnerships with no matter whom. Indeed there are cases where you must preserve a certain distance, outwardly. But in the matter of discernment the individuality is the only one to listen to. I have told you many times when to adopt which attitude, without ever trying to force people to stick together if their natures are irrevocably opposite. No, there are times when it is better to part, to separate, while continuing to maintain good thoughts and feelings for each other. That is much better than to stay together if you detest each other. When you do not know these rules, you always try to arrange things as you wish them, "I like it this way... I don't like that". People use their personal idea of pleasure or displeasure as their guide without realizing what a bad guide they are trusting. It is the quality of reflection, of reasoning that should command our actions, not the fact that something is pleasing or not. All the tragedies in the world come from the fact that our appetites, taste and distaste, our feelings of attraction or repulsion are what control us, never our intelligence.

Now let your personality sit and stew in the corner while you turn to the individuality and ask it to tell you how it feels about things. Since it is high above us it knows everything and can tell you what you want to know. If you never ask it any-thing and go on doing as the personality suggests, you will never solve any of your problems and difficulties. You believe everything you do is for yourself, but if it is the personality's suggestion, it is never for you but for other beings, visible and invisible, who try to influence you because they *gain* some-thing. If you knew you were working for other people's inter-ests you would not be so keen to follow the personality all the time. And that is the point: to know when you are doing something for yourself, and when it is for the benefit of lower entities or family spirits bent on ruining your future. You must do everything in your power to gain control over those entities and reduce them to silence, otherwise you will be in their service like a domestic animal, no more.

Look at animals. Some are fortunate and live in the forest in freedom, but others, poor things... horses, cattle, camels, dogs... work for a master who exploits them. We are not free, either, we are hired by other forces and we work for them at our expense. It is most difficult for humans to understand that their physical body (stomach, intestines, sex) is not really them. We must feed the personality enough to keep it alive, as if it were our horse, our car, but we must not identify with it... it is better to pause and reflect, meditate at all times, in order to know which is urging you on, the personality or the indi-viduality. You have not had that discernment in your life, and the day the light comes to you, you will see that you put all your capital into a bank that was not yours, and now your money has gone down the drain.

Ninth tenths of humanity are slaves, slaves to their hus-band, their wife, their children, their boss, their vice or pas-sion, and if they work only to satisfy these "bosses", they will see sooner or later how little they themselves are satisfied. All that energy has been spent for nothing!

Take sex, for instance. When you give in to a love that is purely sensual, personal, egoistic, your instruments are beyond your control and function independently of you without your being able to stop them or apply the brakes in any way. You notice this, but you are unable to do anything about it. It is obvious that other forces have taken over and do as they like whilst you are there merely as a spectator. In spiritual love it is not that way, you know that it is *you,* your soul, your Spirit, your individuality who are fed and strengthened, and not other forces outside you. No more than a look, a few words, a presence, a whiff of perfume, a strain of music... and you are happy in a way you have never been happy before, because it was you yourself, your higher Self whose thirst was quenched, who breathed these fragrant subtle things and felt joy... not other forces.

Unfortunately most people do not observe themselves, they eat, drink, have "fun", and, as long as their physical body is content and satisfied, they think *they* are. They don't see the emptiness in their soul and Spirit. When you identify with the personality, even if the physical body is sated, sound asleep and snoring, you yourself are starving because the soul and Spirit, the individuality are neglected, starving to death.

In order to understand how true this is, you have to have reached a certain level of evolution. Talk about spiritual love to sensual or primitive people, they will tell you, "But if we don't gratify our sexual needs, we will die! That is what makes us *live.*" Yes, of course, that is what makes the roots live, but the flowers and fruit above are dying. It all depends on the person and his degree of evolution.

I wish you a good afternoon. Spring is approaching, as you see. It won't be long before we will be coming each morning to watch the sunrise. Of course, there is still a month or so to wait, but what is a month to a heart that is loving and patient?

Sèvres, 1 February, 1972

Chapter 11

Identify with the Individuality

"To resuscitate is to have learned to renounce one's weaknesses. That is what we do here and have been doing for years, through prayer and meditation. It is through meditation that we are transformed. Each meditation increases the Light within that builds the Body of the Glory in which one day we will resuscitate!"

If I were to comment on this thought, dear brothers and sisters, I would have too many things to say... for that reason I will stop at the first sentence, "To resuscitate is to have learned to renounce one's weaknesses." A tremendous task, of course... well nigh impossible. We have so many weaknesses, it will take centuries, millenniums to overcome them and get rid of them. However, there is one method that is most effective. Which one? To know yourself. Yes, first of all to know yourself. If you observe yourself you will see that you have two natures, one lower and one higher, the personality and the individuality. I have spoken a great deal about the personality, you should now be able to recognize it when it manifests, and not be fooled by it all the time.

Our weaknesses have their roots in the personality. That is why it is best not to think about them too much; it takes a

whole lifetime to correct just one... and even then it is not
conclusive. Instead, you should try to do something about the
roots themselves, the personality, which is the thing that
keeps your weaknesses alive. The personality, as I have told
you, can be recognized by its egocentricity. A man who abandons himself to his personality thinks only about himself,
without ever seeing the others around him: he takes himself
for the centre of the universe. The world is there, he thinks, to
content him and gratify his every wish, everyone must gravi-
tate around him, treat him with love, wait upon him hand
and foot. Look at lovers: if by chance the boy forgets to look
lovingly at his girl (or vice versa), she is furious: "What! Do
that to me! Not even look at me, not speak to me, not come
and see me!" The fact that he may not have had time, that he
might be tired or ill, is of no importance, she doesn't think
about him, only about herself, and then the quarrels begin,
"Why didn't you come on Sunday, etc...."

As long as his personality occupies the centre, a man will
always be tormented, because he is bound to run into someone who will not accept his demands, who will not consider
him a genius, let alone a divinity with divine rights. All troubles, all human tragedies come from indulging the lower nature to such an extent that it becomes a mountain, blocking
the entrance to the Kingdom of God.

Jesus said it was easier for a camel to pass through the eye
of a needle than for a rich man to enter the Kingdom of God,
but no one has yet explained what this really means. I wanted
to know, and so I searched and searched until I discovered
why. That kind of thing is what I enjoy. People are always
saying, "Enjoy yourself!", and so I do. "Let's see", I said to
myself. "What is the nature of a camel?" And I discovered
that the reason a camel is abstemious to the point of being
able to cross the desert without eating or drinking, is because
his astral body is so small. The astral body of a rich man is
not small but rather extremely large, swollen with desire: he

wants to possess the world! That is why he cannot enter the door of the Kingdom of God, through which pass only those who have overcome their greed. That is what Jesus meant, otherwise it is too stupid. How could a camel pass his huge body through the eye of a needle or a man who is small and insignificant, despite his wealth, not be able to pass through the gates of the Kingdom?

Man's personality is overdeveloped because of the education and upbringing he has received, the advice he has been given, which is always to satisfy the personality, until it has grown into a giant tumour inside most people... you can't say anything at all to them without provoking violent neurotic reactions. People must learn to develop their individuality, forget themselves a bit, enter into other people's situations and say to themselves, "If he didn't come to bring me this or that, maybe he was busy, ill, exhausted..." remaining calm and reasonable instead of immediately planning their revenge. The individuality thinks constantly of others, and that is what I would like to develop in my brothers and sisters.

I will give you an example of the way people usually behave... it may be a bit exaggerated, a bit contrived, but it shows how most people think. Here is a couple, man and wife: now watch what happens. In the morning the husband leaves for work: "Goodbye darling! Goodbye darling!" They kiss each other abstractedly, thinking of other things. As soon as the door is closed, the wife starts to grumble: "Oh, that one, I should never have married him! He is a lazy good-for-nothing, incapable and clumsy... not like our neighbour. Ah! He is someone! They always have the latest car and his wife has beautiful jewelry. Ah! How miserable I am!" She keeps grumbling, "No, I can't stand it any more, when that idiot comes home tonight he will see what he will see!" And she spends the day fulminating, filling herself with venom.

Now let us see what the husband is thinking: "Ah! That b... (I won't say the word) why was I so stupid as to marry her? She is so ordinary, so dumb! All she thinks about is going shopping with her little dog and eating cakes with her girl-friends... while I work here in the dust and noise to bring back the money she will spend on clothes! This cannot go on, she will see what she will see when I come home tonight." Each one grumbles all day long, and when they come together in the evening they tear each other apart. And the same thing the next day, every day.

Now let us see how it *should* be: In the morning before leaving, the husband and wife embrace tenderly, warmly. When he has gone, she says to herself, "Ah, poor darling, when I think what sacrifices he makes for me! How could he have married me, he who is so fine, so noble, so honest! And especially, what love! What a kiss he gave me! He has to work all day under the most difficult conditions, with no time to breathe, whilst I am free all the time, I can rest or take a walk or do anything I like. I am going to cook a wonderful dinner for him tonight." She thinks about him this way, and is happy waiting for him.

In the meanwhile he is thinking, "Ah, why did I make her marry me, poor thing, she's nothing but a victim, all day long she has to clean the house, take care of the children, the washing, the cooking, with no time to do anything for herself... whereas I can have a drink at the pub with my chums, I can talk and laugh with them while she has to wait for the children to come home from school. Really, what a marvellous woman Heaven has given me! I must do something for her." And on the way home he buys some flowers and a little gift to surprise her. They are happy, they bill and coo together with such love!

Now, actually there is no difference between the two couples, neither is better than the other, except that in their *heads*, in their way of looking at things, there is all the difference in the world. It is so easy to change your point of view...

not to change yourself, that is extremely difficult, but to change your point of view. The personality and the individuality are nothing but your point of view.

Do not pay so much attention to the demands of the personality. Even when you have cause to be displeased, you must not react as it suggests, but reason with it, "Listen, if I go along with you and your opinion, I will have nothing but trouble. Why don't you change and become reasonable?" After this little conversation the personality subsides, and that is what should happen, because it must not be allowed to give advice... it should receive advice, but not give it. As no one has a clear idea on this question, everyone goes along with whatever the personality suggests. I have seen it, even the most cultured and learned people let themselves be governed by the personality... believing it is they who make the decisions! No, the personality is like a second skin, but it is not ourselves. Man is the individuality, that is, all that is most intelligent, wise, luminous, immortal, powerful. Yes, that is man! Only he has not yet formed the habit of living in his individuality and identifying with it, he is asleep on that level. Whereas the personality is wide awake, and that is the error, that is where he goes wrong. He must learn to identify with his divine nature.

Initiates in India summed up this work of identification in the formula, "I am He"... that is, only He exists, I do not exist except as a reflection, a shadow of Him. Man cannot exist separately, he is part of God who alone exists, we are a part of what He projects. When we can say, "I am He", we show that we understand we have no existence outside of God, that we are linked to Him and are close to Him and that one day we will be like Him. For thousands of years history has recorded the testimonies of certain beings who were able to identify with God, and who received the divine power, the divine light, ecstasy, joy. When man is ignorant of the reality of his true Self, he identifies with his physical body, his emotions and thoughts, without realizing that they are not reality. That is why he remains forever weak and sickly.

In identifying with the personality you are vulnerable because it will always urge you to expect too much from others, you will never be satisfied. Everyone of us has problems and if you're always waiting for someone else to understand and help you, you will always be unhappy and unsatisfied. Someone will be with you for a moment, and the next moment he or she will be gone. I say to young people, "If you are waiting to be loved, you are going to suffer, because you are counting on something uncertain: one moment you will be loved, and the next moment who knows what will happen? You must not count on the love of others, if it comes, then all the better, let it come and stay forever, but the point is not to count on it. That is why I say, Do you want to be happy? Do not ask to be loved, but *love* night and day, and you will be happy. You may fall in love one day with a wonderful person and share a great love... yes, why not? It can happen, but don't count on it." I have solved the problem by counting only on my own love. I wish to love, and if others do not wish to, that is their affair, they will be unhappy as a result, and I will be happy! Problem solved. If you find a better solution let me know.

The attitude you should have toward the personality can be summarized in one or two sentences. Firstly, do not identify with it, but identify with your divine Self, repeating the formula, "I am He". Secondly, do not declare war on one weakness at a time, but on the entire personality, since that is what is responsible for your faults and weaknesses, and keeps feeding them. I have never seen a more prolific woman than the personality, you have no idea how prolific! It produces without cease, but the children are monsters. And thirdly, instead of giving in to the promptings of the personality, watch over yourself constantly, keep yourself in check, keep a firm rein on the personality.

The personality is responsible for most of the
which are commited. Instead of letting it be the mistres
house, the ruler of your kingdom, who governs you a... tens
you what to do, make it into a servant, and then you will obtain everything you want, because it is indefatigable and well-
armed... yes, with claws, fangs, hooves, horns, forked tongue
and all! A whole arsenal. And it knows how to scratch, especially in women, it scratches and bites, and pulls hair!
Whereas men give blows. Yes, a woman's personality is
armed a bit differently than man's.

I have studied the personality so thoroughly that now I believe I know it extremely well, how it walks, how it talks, how
it laughs, how it eats, what advice it gives. Really, it is a whole
world in itself, the personality! You also should study it, get
to know its gestures, its language, its look, its colours. The
colours of the personality are never luminous, never radiant,
except perhaps in the midst of sex. Then it lights up for a moment, but the light does not last, very quickly it becomes dull
again. And when it is angry, it gives such a black look, as
black as pitch! Instantly you know it is the personality looking at you. Of course the personality is also capable of kissing
and caressing, but always in order to get something. When the
individuality kisses and caresses it brings you all the music
and poetry of Heaven. Both can embrace, but the difference
in their kisses is something you have no doubt never noticed.
Do you know, when someone kisses you, whether it is his personality or his individuality?

I can give you a criterion so that you will always know.
When it is the personality, it is like the suction of a leech, an
octopus, you become exhausted. It benefits, whilst you grow
poorer, you are the loser, you are demagnetized... it robs you
of everything you have. The individuality has the opposite effect, you feel richer and better, more expansive. A new light
on this subject! When the personality kisses you it wants to
take everything from you without a thought of what state you

will be in, it cares about enjoying its feast, not about you. Whereas the individuality wants to give you what it has in its heart and soul, as a result of which for days you feel enriched, embellished, overjoyed. This is what young people should learn, to discern when it is the personality which is manifesting during what they call love.

So, dear brothers and sisters, analyze yourselves, observe what state you are in during the day, all day long, and when you see that the personality is becoming too forward, say, "Be quiet, you, otherwise I'll starve you to death!" Yes, you must threaten it and make it listen, otherwise you will be lulled by its persuasive reasoning... until you are lost. Listen to the reasoning of the individuality. You say, "But how does the individuality reason?" Oh! The individuality reasons beautifully. Whatever happens it always sees the bright side. Faced with difficulties, illness, accidents, it believes that good will come from that evil, and that is why everything always works out! It says, "What is a little sorrow if in the end it means happiness?" That is the right way to reason.

I will give you another example. You may find it a bit contrived perhaps, like the other story earlier, but never mind, it will make you understand something. Picture a young man out of work. Wherever he goes he is refused. He is nonetheless capable, honest, loyal... yes, but humans are often blind. One evening, after still another refusal, he is walking aimlessly and sadly in the street when all of a sudden a car with an idiot at the wheel runs him down. The idiot speeds away, leaving him unconscious in the street. Calamity! Yes, but is it really such a calamity? A few minutes later a millionaire comes by in his car, and, seeing the young man lying prone, being kind and good-hearted he carries him to his car, and drives him home. Naturally the millionaire has a daughter who undertakes to care for the young man and nurse him back to health. Will it surprise you to hear that they fall in love with each other?

The father who sees nothing wrong with the situation, gi
them his blessing and they marry and live happily ever aft .
The young man also inherits the fortune of the kind millionaire and, as you see, this calamity led to happiness!

You say, "But it's only a story!" No, there are many such
cases. It may not be exactly as I told it, but under one form or
another it often happens that, thanks to some tragedy, people
fall into something wonderful. If they had not undergone set-
backs and difficulties, they would never have succeeded in doing anything grand, or noble, or divine. That is the way to
reason when you find yourself in difficulties and that is precisely the reasoning of the individuality.

So now, get down to work! Survey yourself, analyse yourself. This doesn't mean that in a few minutes you will be
transformed, but when your personality sees that you are becoming strong and powerful enough to command and dominate, it will subside into the back seat, where it belongs. This
subject must be understood now, for it is the key to the problems of life. Those who do, obtain results: in many of life's
most difficult circumstances, when others around them are
capitulating, they know what to do.

Sèvres, 3 April, 1972

Chapter 12

The True Meaning of Sacrifice

Commentary after the lecture of 28 August, 1971: *The Tree.* "From the level of the animal kingdom all creatures undergo suffering at being eaten by another creature. For those who are able to lift themselves into the celestial regions the pain disappears and is transformed into joy, for to be devoured by the heavenly Beings is a joy, a felicity that words cannot describe. It is in this sense that all religions have advised man to offer himself to God in holocaust... so that God might feast upon him."

It shocked you when you first heard that the Lord enjoys devouring His creatures, the idea had never been presented that way before! I know, I know: Christians believe in a God who never eats or drinks, never breathes, who in fact has no needs. But I discovered on the contrary, that He enjoys feasting! This is normal is it not, since we were created in His image and we eat? He also must eat. Otherwise, when we eat, whom do we resemble? God nourishes Himself, He eats the best kind of food furnished by the holy creatures nearest to Him who are pure light and pure love, called Seraphim, Cherubim and Thrones in both the Initiatic Tradition and in

Christianity. I have already explained what these creatures
are like, how they manifest and what their qualities and virtues, their colours are.

Throughout Nature you will find that all creatures serve
as nourishment for other creatures, this is the law. If you are
not strong and resistant, then it will be evil that eats you; if
you are prepared, then it is you who eat evil, that is to say,
you make it retreat and diminish, you weaken it until it vanishes... and then you benefit from its energy and substance.
The psychology of the future is based on our understanding of
this truth. Do you remember in the Talmud it is written that
at the end of time, the saints, the elite of God, will be fed the
meat of the Leviathan, the sea-monster which represents the
Devil. The Leviathan will be cut up, salted and seasoned, prepared by the Lord Himself (perhaps He has already put it in a
deep-freeze to preserve it) and placed before the Just. What a
feast awaits us! But this is symbolic; literally it is too horrible
that God should give us a monster to eat! Those who understand symbols know that in reality it is all a question of particles, energies and forces which man absorbs, in which case
even evil can be nourishing, on condition that we know how
to dilute it and absorb it in homeopathic doses.

Evil is no more than condensed energy, very potent energy, which the human system cannot take because it is so
concentrated, but in homeopathic doses this energy becomes
a most effective remedy. As I told you one day, the venom of
a cobra would cure a number of diseases if we knew how to
prepare it and divide it into doses. Yes, poisons can be useful
and have already been used by the medical world. Dangerous
venimous plants which have killed so many ignorant people,
and which sorcerers and black magicians used in the past for
their diabolic purposes, and even the violent drugs used by
the youth of today, will one day be used for good, for health.
Everything that exists in Nature can be used for health, for
balance and for life. If man is not prepared, if he is too weak,

too ignorant, he succumbs and lets himself be devoured. Because evil also needs to eat! And so it eats humans, it devours them in all kinds of ways: illness, vice, anguish, etc... but if they become very strong then it is they who will eat evil. Like the Initiates... things that make other people succumb, difficulties, misfortune, accidents, only make them stronger because they force themselves to find the energy and the means, and become supermen. What is the difference between an Initiate and an ordinary man? The way they consider evil. Unless man is told what his possibilities are, evil will always be harmful and dangerous, but as soon as the light comes, evil will be something he can feast on.

Why is sacrifice of such tremendous importance in all religions? Why did they immolate animals, sometimes humans, in the past? Why was it important to make frequent sacrifices to the gods, to Jehovah? Why does it say in the Bible that the smoke from the sacrifices was a pleasing odour to the Lord? What was the meaning behind the sacrifices? And why, with the coming of Jesus, did it all change? From then on it was no longer calves and lambs that man was asked to sacrifice, but his inner animals, that is to say, his weakness and passion, his greed and lust, his sensuality. That is the true sacrifice, to give up one's instincts and transform the brute force within into pure, luminous, sublime energy.

At the ceremony of the fire, when all those twisted, blackened, ugly branches burn and turn into light and heat, I have explained that that was the true meaning of sacrifice. Those who refuse to understand and who do not wish to offer themselves in sacrifice, as food for the Lord, remain as they are: animals, insects, monsters; those who ask to be consumed in the fire, the divine Fire of divine Love, not only do not die, they resuscitate! That is the meaning behind the words of Jesus: "Unless ye die ye shall not live." We must die, but in what way, how? Not, obviously, wounded to death by a knife or shot by a revolver. Jesus was not referring to physical death, he was talking about dying to the personality, to everything on the lower plane such as desire, vice, passion... in order to come alive on the higher plane of the individuality, there where it is really nourished.

Usually it is the personality who feeds on us, its greatest wish being to seize and devour us. Twenty, thirty, fifty times a day it grabs us and feeds on us! The result is that we grow weaker and weaker whilst it grows stronger and stronger and more and more resistant. But if we learn how to summon the individuality to our aid (it too is hungry and knows very well how to go about overcoming the personality), soon nothing will be left of the personality.

Everything in life eats. Look at the way iron is eaten by rust! Yes, everything eats everything else. People say, "It is either him or me... slay or be slain", meaning, if I don't eat him, he will devour me. Good eats evil, and evil eats good. We must let ourselves be eaten, but only by our higher Self... then instead of suffering we will rejoice!

In Tibet, there are certain Initiatic rites that require the disciple or candidate to spend the night alone in some terrifying desert place, alone, where he must challenge the demonic spirits to come and devour him. It is a terrible test which often finishes tragically, many cannot stand it and go mad, or die. I do not agree with this kind of thing. I think we should ask the Lord to send us Angels, to ask Angels to eat us, because they have such a developed sense of beauty, of measure, of wisdom, that they will first of all subdue the personality and set us free. If you are fearful of asking to be eaten by the Angels, you will surely die. You must pass through death in order to live, really live, which is what is meant by dying in order to live, or: "Unless ye die ye shall not live."

Death was the main preoccupation of the Teachings in the past. The Egyptian religion, for instance, was a philosophy of death, concerned only with death and the beyond, and the greatest of Egyptian gods was Osiris, the god of Death. In the

Initiations the ultimate test was that of the tomb: only the man who accepted to die could hope to be resuscitated.

You have no doubt read how Socrates died. It was a noble death. Socrates had been learning how to die during his entire life, and that is why he was able to accept his condemnation with equanimity and serenity. His death is a great example to this day. But in the Western World, especially now at this time in history, people are afraid of death. In India when a man dies, he is incinerated, and it has been the tradition for his wife to throw herself into the same fire. I will not go into whether this is good or bad, stupid, useless and cruel, or not. I will simply state it as a fact that this custom existed for definite reasons: like all customs, it was founded on the knowledge of certain laws, created to teach man that he must conquer fear. And the fear of death is the most powerful, the fear of dying of hunger, in poverty and misery, in deprivation....

Today I ask you to retain one thing especially, that good and evil devour each other mutually. If you let good predominate, it knows very well how to swallow up evil, but if you let evil have its way, if only for an instant, it swallows up the good immediately. It is the same for the individuality: put it in first place, you will see how the personality is immediately weakened, gnawed and devoured; but give precedence to the personality and it is you who become thin, weak and pale... and fearful.

Everything in life eats something else. The husband eats his wife and the wife eats her husband. Sometimes the wife's health declines because the husband devours her or sucks her like a vampire. Or else it is the husband who weakens, who loses his health because his wife, without knowing it, takes all his strength. These are universal laws, everything is planned for the purpose of one creature being nourished by another, but of course there are limits! Knowing that, if you really nt your personality to be eaten, which will give you the greatest possible advantages, summon the individuality and say, "Do something about this damned personality, it is torturing me, it is too much for me, do something about it!" And then, let it do as it wishes.

If there is one subject that interests and absorbs me above all others, it is the question of the personality and the individuality, for if we could ever see these things clearly, it would avoid a great many catastrophes, tragedies. From now on, put aside everything else, if you will, and concentrate only on these two factors: you will see how everything will become clear and light-filled for you. I have spent my life training myself, doing so many exercises that I know at every moment of the day or night whether it is the individuality or the personality which is manifesting through me. When I am contemplating doing something, there is a little clicking sound that goes off inside me, an inner computer as it were, that informs me. I would like you to get to the point of having the same awareness, the same lucidity all the time, because on it depends your whole future. You must exert yourself, train yourself as I did, to let nothing pass through you without identifying it. Then whether you choose the right way or not is another question. Perhaps you will want to go along with the personality, who knows? But the important thing is to identify it and to know quite clearly which it is you are going along with. It is better to know where you are, and then decide what you want to do... but first of all to know. When you love someone for instance, you must immediately try to know whether it is your divine nature which is manifesting your love, or simply the personality once again taking, devouring, grasping... satisfying itself at the other person's expense without a thought for his future, like a blood-sucking vampire.

I am exaggerating a bit by insisting so much, but if there is no one to repeat and emphasize the important things you will stop trying, you will give up. If I stop, you will never go on by yourselves. You only want one thing really, which is to be left in peace, but I am not going to leave you in peace, on the contrary, I want to show you how far you still have to go. That is why I keep insisting, and, if I bother you to death or bore you to tears, all the better for you!

Le Bonfin, 5 August, 1972

Chapter 13

The Balance Restored

I have observed that most people, even the most learned and cultured, do not hesitate to put themselves at the disposal of their personality. Without realizing it they give their most noble faculties, their intelligence, creativity, sensitivity and all their time to satisfying their lower desires. It never occurs to them that these faculties should be put to work to help accomplish something grandiose, something important, such as realizing a High Ideal.

The Kingdom of God will not come on earth until all mankind gathers together the forces of the higher nature and puts them to work toward this Ideal. At the moment human beings are not enlightened, they give all their energy, even their soul and Spirit, to satisfy their lower, coarse side, the personality. And the strange thing is that the more they try to content that side, the more empty and dissatisfied they become... their lower nature is sated, full to overflowing, overfed, while the higher nature suffers from neglect and starvation, it is wasting away.

I receive many letters from brothers and sisters saying, "Oh, Master, before I knew the truth, before I knew about the personality and the individuality, I did everything to gratify my personality, and the more I tried, the more unsatisfied I was... and I blamed myself." That is what happens when you leave the higher nature without food, it pesters us with reproaches, "So, never anything for me?" That is what causes your dissatisfaction.

No other era has offered mankind as much with which to content the personality, and yet never has man been less contented. In the past he had very little, practically nothing, now everything is at his beck and call, but with it all, people are unhappy, empty, unbalanced, disturbed, nervous, unhinged. Our technology, inventions and discoveries are all put to work for the benefit of the lower nature, which is overfed to the point of being sick. Why do human beings not understand that they have other needs which need to be fed? It is unbelievable that in this twentieth century, the century of light supposedly, people have not seen the essential. The more they have, the more they lack. Like the story of the husband who gave his wife everything she could possibly want, clothes, jewels, cars, villas, all except one thing: he forgot the essential... to love her. And so, as she was not happy, one day she ran away with the chauffeur! Doubtless the chauffeur succeeded in giving her the love she was missing, and she abandoned everything else for him. As long as you deny someone the subtle food he needs for his soul and Spirit, no matter what else you do for him or her, one day you will have a surprise... you will be abandoned.

A wife comes and tells me, "I have done everything for my husband, not only have I surrounded him with love and attention, but I tried in every way to satisfy his slightest desire... yet he left me!" "Ah," I answer. "For whom?" "Well, that's the trouble, he has gone off with my best friend who is cold as ice!" "That's it," I say. "You were too hot, now he has gone to cool off a bit!" And it is true, many women do everything to content their husbands in the realm of the personality, his stomach and sex, but they are unable to awaken in him anything higher, more divine, and so the poor thing has to look elsewhere. I am also aware that there are plenty of gross and vulgar husbands, but that is another question.

You already know everything I am saying, this is not the first time you are hearing it. But what I would like to add today on the subject of the personality and the individuality has to do with the exercises we do every morning. These movements are so simple, so easy, everyone can do them. What a contrast with the gymnastics that require great physical effort! People are convinced that to be strong and healthy they must have tremendous muscles. No! I have often told you that even if your muscles are strong as Hercules or Tarzan, if your nervous system does not function properly you will be so flabby and so soft-witted that you won't be able to pick up a feather, simply because certain currents do not reach the muscles to stimulate them. In asylums you see inmates who become so strong during a fit that it takes four guards to overcome them! It is because of a current which descends from the brain and circulates through the muscles so violently sometimes that it contracts them. The nervous system is extremely important, as you see, and the exercises we do here, although they may not develop the muscles, do reinforce and harmonize the nervous system.

I will go very quickly over what each movement does for you, and then I will stop and explain the fifth movement, which concerns the individuality and the personality. The first movement teaches us to receive the forces of Heaven, to bring them down into us so that our cells may be cleansed and purified. The second movement is to bring the magnetic current of the earth inside us: the union of the two currents, Heaven and earth, fills the solar plexus with perfect harmony. The third movement teaches us to swim in the ocean of cosmic Light. The fourth teaches us how to sever our bonds with evil, the evil that binds us and ties us down. The fifth movement is to teach us how to remain in balance. And with the sixth, we chase away all the impurities we have accumulated.

But it is the fifth movement that I wish to talk about to-day. If I do not explain it, you will never know what it means, nor how important it is. When you do these movements to the left, to the right, with your arms bent and your hands resting on your shoulders, have you noticed that the upper part of the body must do certain *other* movements in order to preserve the equilibrium? Look at my way of reasoning: this is exactly what happens in life. When you release some movement in the lower nature (a thought, feeling or desire), as everything in man is related, it may create a disorder elsewhere, an imbalance occurs *unless* the higher Self is there to supervise and control. Acrobats on the high wire must constantly keep their balance with counter movements of the arms, otherwise they would fall and break their necks. You have all watched a tight-rope walker at one time or another but you have drawn no conclusions from what you saw.

Of course your equilibrium is not so much a question of the conscious mind as the functioning of something we all have in the ear. In the inner ear, there is a centre in which are tiny semi-circular channels or ducts, filled with a liquid in which float tiny crystals. Along the walls of these ducts or canals are ciliated cells, articulated by nerves which transmit messages to the brain extremely rapidly. If the nerves do not function properly, you lose your balance immediately. Acrobats who have no time to reflect on what action they must take, train themselves in such a way that they react instantly. Instinctively the acrobat then does what should be done.

When you are driving a car there are so many unexpected emergencies to contend with that if you took the time to think about what to do each time you would have countless accidents. Good drivers avoid danger because of their quick reflexes: it is the same apparatus functioning. The Initiates say that the ear is linked with wisdom, because our whole balance depends upon this apparatus within the ear. Worries, disharmony, trouble, problems, are the cause of countless accidents.

That is why it is so important when you are driving to remain calm, clear-headed, aware and reasonable; only when your mind is free can you react instinctively, automatically, with great rapidity, to avoid an accident. Even experienced and adroit drivers notice that when they are anxious about something their reflexes work more slowly and they are at the mercy of circumstance.

Therefore, in the fifth exercise, as you move your legs to the left and right, the upper part of the body must counterbalance the movement of the lower part, it must adjust to maintain the equilibrium. How many lessons to be learned from that one fact! But people are not used to seeing things in life as they are, they look at them without seeing, thus missing their profound meaning; they put what they don't understand into a back drawer and get on with their ordinary lives. A husband who is planning to change his life radically for some business or other reason, must consult his higher Self before deciding, otherwise he runs the risk of having all kinds of trouble he has not foreseen. His wife and children will be affected, they must also be considered. How many people I have seen make decisions without taking other people into consideration, without foreseeing the consequences or taking precautions to avoid tragedy. Every time one makes a change without foreseeing the consequences there are repercussions, regrettable repercussions. The point is to foresee, to be aware of what might happen.

A small innovation may cause tremendous upheavals in other realms, economic, political, scientific, technical, religious, etc.... Take the example of the car: every single area of men's lives was affected and completely changed by this one invention! The rhythm of life, in government, business, the law, in work or leisure, not to mention hygiene and pollution, diseases of the heart and nervous diseases, are all results of the invention of the car. And what changes it wrought in homes, in families! The future of a marriage can hang on a car... either because you have none, or perhaps because there are two... and how many liaisons or broken relationships occur as the result of the family car?

Now, what is the personality? It is the movement, the action of the lower nature. If we do not appeal to the individuality to react, to maintain balance, we fall, we become victims of accidents and all kinds of trouble. It happens all the time, every day you see more and more people who are unbalanced, disoriented, lost, because of indulging the lower nature, its appetites, its craving for pleasure, its caprices and whims, its lust... placing it above the individuality always ends in trouble. A man who sits down to eat in a restaurant and orders without checking the prices, when the waiter arrives with the bill, with find the amount astronomical! If he lacks sufficient money with which to pay, he is carted off to the nearest gaol... symbolically. This is what happens to all who listen exclusively to the personality.

My dear brothers and sisters, if you accept these great truths you will avoid a great deal of trouble, you will see precisely where you stand, where you went wrong, and see the future clearly. You will start your life over again. That is worth billions! Unfortunately, the catastrophic advice of the personality tells you to keep a firm grip on your worries and problems even when you are here, in Paradise. Why? To keep you from learning, of course. The personality knows that when the light comes, it will have to disappear... that is why it does not want to see us climb out of our ignorance. How often I see people arriving here with their burdens! I say, "Set those down at the gate and come and learn some great truths. Then when you leave, you can pick them up again if you must." But no, they hold on to them, they don't want to be separated from their problems! Certain brothers

and sisters never make any progress and I see why : they don't
realize how badly they are influenced by the personality.

And wait, there are still more pitfalls that you don't know
about. This is what the personality says to you : "Poor old fellow, you are dust and to dust you will return, so who do you
think you are? You must be good, you must be reasonable
and conform to what the crowd thinks, there is no other
way." The personality persuades you that you cannot escape
mediocrity, and then you react like a hypnotized hen around
which somebody has traced a chalk circle and, as it thinks it
cannot step out, it does not even try! Man is like the hen ; the
personality keeps telling him, "You are limited, you are narrow, you are mortal, you can't go further, you must restrict
your ambitions, you must stay within your limitations..." and
so the poor thing is bewitched and doesn't move... until finally the individuality steps in and says, "But that is only a chalk
line! You are free, you can pass, go on, go ahead!" You try,
and lo and behold: you can! The individuality is the one to
listen to, because it has no limits, it is always encouraging,
"Go on, you can do it, you have enough strength and power
to go all the way to infinity if you want to!"

Retain this one thing from today's talk : the personality
sets traps for you, once you know that, you need not let yourself be influenced any more by the worries and fears that try
to invade you... that may follow you even to this Divine
School. Your attention should not be diverted into negative
states, how do you expect the divine side to filter in, if it is?
People come here and understand nothing, nothing at all...
and yet actually it is so easy to understand! They bar their
own way. We must learn the truth in order to stop this feeling
of being limited.

How many more things I could reveal to you about the
personality, its ways, its clever manipulations that fool us
every time because we lack understanding, we lack the light. I
have been given this role of throwing light on the reality of

things, so that you will advance, so that you will evolve. But it is as though you did not want to progress, you cling to the old prejudices. If I cannot help you, even with the Light of this Teaching, who will? Perhaps you think God will. Yes, but you know God is not terribly concerned with details, He has other more important things to think about. He confides us to His faithful servants, to whom we should listen! And if we do not listen, what can He do? Yes, war, illness and disease, accidents, all come from not listening to them. You say, "Why does God not come and take us out of all that?" I am not saying He never helps us, but what can He do, even He, if we are closed?

Take the sun. The sun is powerful enough to make the planets revolve, it moves them and vivifies them... and yet, if you close the shutters of your windows and draw the curtains, in spite of all its power the sun cannot enter. We leave the shutters closed and then we say, "Come in, come in, dear sun: I invite you to enter!" And the sun answers, "I cannot!" "Why not?" "Because of the shutters!" Whoever understands what I am saying will open their shutters and the sun will enter and pour light on him. The sun is the symbol of Almighty God. God is all powerful, He is responsible for the whole universe and all existence, but when it is a question of opening the shutters, He cannot: it is up to us to let Him in.

Some people who are good Christians, firm believers, think they should abandon themselves completely into God's hands. Then why is it that God leaves them alone in the midst of their difficulties? Because they don't know what to do in order to attract His blessings. They think it is enough to placate Him, the way they invite someone in a high position to dinner in order to obtain a favour, a special authorisation. Many people think they can do the same with God. If they cannot invite Him to dinner, they promise to light a candle for Him... as if one could buy the Lord with a candle! He *can* be won, yes, but in quite another way. You must discover what His Nature is like, discover His tastes (might He not have certain preferences, the Lord?) and do your best to please Him. Then He will grant what you ask immediately.

I know some of the things the Lord likes. He appreciates enormously a quality called gratitude, for instance... if someone is grateful He relents! Like a father. A father asks nothing from his child, he gives him everything without asking for anything in return... but one thing makes him happy: to have his son recognize his goodness and generosity. Otherwise it makes him... not angry, but still a little annoyed. God is that way too. He needs nothing from us, He has everything, but He loves to see that His children recognize that He is a good Father.

I know another preference of the Lord: He loves us to act unselfishly, with impersonal detachment. When someone gives Him all he has and consecrates his life to Him saying, "Here, Lord, is everything I have, it is all Yours to do with as You wish..." then God surrenders! But if you threaten not to go to church any more, if you ask Him to make your wife die so that you can marry someone else... this surprises you? If you knew many such prayers Heaven receives! "Lord God! Make my husband die so that I can marry my lover...." Well, Heaven does not even hear that prayer. There are too many prayers asking for money, for cars, for material possessions, and so in Heaven they say to each other (I have heard them!), "Oh, my! Nothing but selfish requests! We are overburdened, up to our eyes and exhausted by all the letters requesting money, pleasure and ease, women, children, diplomas and degrees. They will have to wait, for it will take a long time to go through all this paperwork!" But as soon as they get a request to serve God.... Ah! this is so rare and so highly appreciated, right away they are delighted and do everything to grant the request.

You see how well-informed I am? If you do not believe me, go and see for yourselves. They will tell you, "We never grant selfish prayers, we are snowed under by that kind." For thoughts are letters! But you must not forget to enclose a stamp, otherwise your prayer will not be fulfilled! The stamp is symbolic, of course. And it is up to you to discover what stamp to use.

Le Bonfin, 8 August, 1972

Chapter 14

Render Therefore Unto Caesar...

"When you give in to the personality, you think you are doing it for yourself, in your own best interests, and that is why you are so persistent. If you realized that it is not for you but for beings, both visible and invisible, who are urging you to work for *their* interests, not yours, you would not be so eager and certain. What you must learn is when what you are doing is for you, yourself, and when it is for the benefit of creatures who have nothing to do with you."

The subtle question of when we are ourselves, and when we are not ourselves, is what I wish to throw light on. For therein lies the tragedy. People make the mistake of thinking their physical body is themselves and everything they do is for it, while their soul and Spirit are left unfed, uncared for, starving to death. There is virtually nothing human beings cannot contrive to give to the physical body! And then it surprises them that they are unhappy. Happiness does not depend on the physical body, that is where they go wrong. The physical body is easy to satisfy, a little food, a little clothing, a bed, contents it. But all the food and clothing and beds in the world will never content the soul and Spirit! That confusion is what creates tragedy. We must understand that we are made of several principles and that the needs of one principle are not the needs of another.

I am not saying we must let the personality die, no, we must give it a little something from time to time, but the point is to know how much. To answer that question I remind you of the scene in which the Pharisees asked Jesus how much to give Caesar in the way of taxes, hoping that his answer would provide them with the excuse to arrest him, and Jesus, reading their thoughts, answered, "Give me a coin." They gave him one. "Whose image is this?" he asked. "Caesar's," they answered. "Very well, then render unto Caesar the things which are Caesar's, and unto God the things which are God's".

The answer is well known, and yet during the two thousand years people have been quoting it, no one has ever interpreted it properly. I tried to penetrate into Jesus' thoughts at the moment he was voicing them, and what I found was that Caesar was nothing but another form of the personality. Yes, we all have a "Caesar" constantly demanding its rights, and we must give it something, but not all! You say, "Well then, how much?"

To make it perfectly clear, I will ask you to imagine that you are burning a piece of wood, the branch of a tree... what do you see, what is the result? First there are flames bursting forth as they do at the fire of Michaelmas, and then a little gas emerges, and then steam or vapour, and in the end, all that is left is a little pile of ashes. Where did they go, the flames, the gas, the steam? They went heavenward, while the ashes remained below, on earth. That shows what you should give to the personality: one quarter of the whole, the part that corresponds with earth, and three quarters to the individuality or heaven. Yes, one quarter is enough for the personality, to keep it alive, to prevent it from dying, but all the rest should be given to the individuality.

In another talk I told you, "As the individuality is situated very high above, it is able to inform you, it can tell you anything you wish to know." This is not clear; here is a picture to clarify it. A great scientist or scholar, with degrees from several universities, sits working at his desk, in his study or laboratory on the ground floor, whilst his son aged twelve, is out of doors climbing trees. The child has no degree, no diploma, but *from his position in the tree tops*, he can see far into the distance. "Papa!" he cries. "Here come my aunt and uncle!" And the father who sees nothing, asks the child, "Are they in a car or on foot? How far off are they? Who is with them?" And the child, with no intellectual faculties whatsoever can see better and further than the brilliant older man who is his father! Why? Simply because the scientist is on the ground floor and the youngster has climbed to the heights.

It is the same for us. If we remove ourselves from this personality which knows everything supposedly, but which actually sees nothing, if we climb high above to the level of the individuality, we will be able to see clearly and correctly even without a scholastic background. I am like this child (what vanity!) in that I see what scientists do not see: someone has placed me high up in a "tree", above those who are on the ground. If you remain attached to the personality and consecrate all your faculties exclusively to catering to it, you will see that the results are not splendid. This is true in the realm of knowledge and it is true in the realm of love: everyone seeks love too low down, demonstrated through the personality instead of seeking love higher up, demonstrated through the individuality.

Why do you always identify with the physical body... do you know what the result will be? You are creating your own vulnerability, weakness, illness, death. The body is perishable, vulnerable, ephemeral, which is what you become when you identify with it. Why not identify yourself with something that does not die, that is eternal, incorruptible, unchange-able... which will make you eternal? They say mankind is threatened with disappearance, why? Because everyone iden-tifies with the physical body, without an idea of the conse-quences. Initiates do not identify with the physical body, they think, "It is my horse, my donkey, I will feed it from time to time, but I myself am not that: I am the rider!" People who are ignorant think they are the horse and leave out the rider... their horse is riderless.

Le Bonfin, 10 August, 1972

Chapter 15

The New Philosophy

"The personality is so tough, no matter what you do with it, cook it, fry it, boil it... it is still there; from the bottom of the pot it will look up at you and say, 'Hello! Here I am!' It is really extraordinary!"

Yes, because it never changes, it is constantly linked with the lower regions and all its manifestations are but reflections of those regions. The only way to bring about a change is to replace the personality with the individuality and, once you do, it will no longer be the personality manifesting itself, but the individuality.

Today to make things a little clearer, I will add a word or two. The physical body is the personality's last hope, it is fighting with its back to the wall, and will continue to the end to manifest egocentric thoughts and feelings from its centre of activity, the mind and heart, the physical body. The physical body does not change, no matter what. A man may experience enlightenment, illumination, ecstasy, but that will not change his nose, mouth, intestines, sex... the body does not change.

Imagine a lead pipe through which you pour dirty water, clean water, petrol, wine or anything else: the pipe does not change, it remains what it is and always was. Or imagine a telephone booth: there are different people in it all the time, coming and going, but the cabin does not change. Our physical bodies are like that, they remain as they are, tall, little, hunchbacked, etc... and no matter how upright, intelligent, kind and good the personality shows itself to be on occasion, it will still always be the personality.

That is why I am asking you today to give up all hope of being able to transform your personality. And do not reproach me for my apparent cruelty, in this case it will do you good to have me take away your faith, hope and love for this rickety character... yes, rickety. You stand it up on its two legs and it falls right back down. I would never tamper with your faith and hope in your higher Self, because it is always steadfast, faithful and true. Unfortunately for them, a lot of people are fooled by those two natures. How many wives, thinking to hold on to their husband, have nourished his appetites, sexual and otherwise, and are then surprised when he is unfaithful, ungrateful... and abandons them! Well, those wives are dull and stupid, they should not have confined themselves to feeding their husband's personality only, because by nature the personality is unfaithful and ungrateful, it forgets immediately what one does for it. No one should be surprised at any show of ingratitude or coarseness under those conditions, it is their fault for not awakening the higher nature which would have brought both of them nothing but happiness!

My principle is precisely *not* to satisfy your personality. This makes you annoyed and discontented with me, does it not? Too bad. What I want is to feed your spirit, your divine side which is lying in a corner half dead because no one takes care of it. You are always expecting me to take care of your personality, to flatter you and compliment you... as they do in the world... well no! That would do nothing for your evolution. On the other hand, if I do something for your individuality, I will be bringing you untold riches, I will be helping you to evolve. You will remember this for centuries and millenniums and come looking for me amid the stars to thank me, because the individuality is loyal, the individuality is grateful!

You see, we are so ignorant about so many things, and unfortunately we seem to want to remain that way. Everyone here could be learning how to deal with their problems and how to solve any future ones, instead of which they remain indifferent. There is no money involved for them, you see. This great wisdom will not furnish them with food and drink, nor sums of money, nor amusement. And that is all humans generally seek. To learn something really important doesn't appeal to them... they don't even recognize the value. But I know, I have a knowledge of what is valuable and what is not, in fact that is what interests me most. Therefore from now on, know that it is the individuality, your Spirit, that interests me, I am working on it all the time in order to help liberate you. Even if you are angry I will go on doing it, thinking to myself, "The day they understand, they will stop being displeased."

I have compared the personality to a water pipe and a telephone booth but it also resembles a loudspeaker. A loudspeaker does not change with each different programme, each strain of music; in the same way the individuality manifests itself through the loudspeaker, the personality, without affecting it. You must realize that even if the personality seems to be doing something wonderful, it has not changed, not improved, it is the other, higher nature, the individuality, manifesting through it. No matter how many ornaments you deck the loudspeaker with, no matter how many coats of paint it receives, it does not change basically. Someone may change his physical appearance with facial surgery, but his personality will remain the same, as selfish, ignorant, capricious as ever. When a man sees the individuality, the Spirit, the divine side manifesting through him, he thinks his personality is no longer the same... but as soon as the individuality subsides, there it is, large as life, with all the same inferior tendencies. Are you beginning to understand?

We human beings, in between the personality and the individuality, are responsible for both manifestations. If we summon the individuality, it will come; if we do not, it will be the personality which manifests. You say, "But then, who are 'we'?" A screen, we are the screen on which all kinds of forms are reflected, ugly or beautiful, shadowy or brilliant, good or bad.

You say, "It seems the more deeply we probe into this Teaching, the more disturbed we become, we have the impression of not understanding anything anymore." I know, I know, that is the way for a while, but it is only temporary. Many who come here were much better off before, they were satisfied as they were and everything went well for them. Now, since they decided to change their lives because they became enthusiastic and inspired by the Teaching and want to live a divine life ... nothing works any more, you can see them fermenting.

Of course, to choose the right path does not mean that everything will immediately be perfect! Because people do not know how to interpret things, they are tempted to go back to the way they were: "I was much better before, look at all the trouble I'm having now!" they lament. But that is because they are cleaning their house, everything is being swept away, and so of course the spiders, bats, toads and rats come to the surface (when you tear down an old house that is always what happens), and they must be chased away. But it does not mean that the actual cleansing process is bad, on the contrary, with a little patience you will see, everything will soon be calm and orderly, filled with light.

It is up to you to put your ideas in tune with mine and not the reverse, as you wish it to be. You compare my ideas with yours, and you think, "Ah, no! That is not right, what he means is...." You hang on to your ideas, your points of view, and you reject whatever does not suit you. Well, you should do the opposite. If you want to evolve, you must now for the first time in your life decide to reject a great many of your antiquated ideas, and conform once and for all to this philosophy. Many come to this Teaching to take a look at it, to analyze it and pick out what suits them, thinking, "Ah, this is magnificent!" because it coincides with their way of thinking. They leave the rest aside because it does not appeal to them. We should do the opposite. I did. Ever since I was very young, I wanted to adapt, to conform, to shape myself along the line of a better philosophy than my own, the philosophy of the Initiates. But most humans take what they fancy and give it their approval; whatever they do not fancy they leave aside. Well no, all of you should accept a great many things that do not suit you or appeal to you, because they are *better,* above and beyond what you have known. I see so many people following the impulses given them by the personality, they confine themselves to whatever gives them pleasure. Well, that is not the way to evolve.

I have seen it so often! People will not budge one iota from their own convictions, their own appetites, not realizing that to advance, to grow, to evolve and become perfect, we must get rid of all our personal tendencies. Ah, no! We protect them, we reinforce them and even fight for them as if our salvation depended on them. The whole world does this, everyone is ready to fight for their mouldy old ideas. Well, I determine someone's grandeur by whether or not he decides to sweep all his ancient concepts away in exchange for a new philosophy, this divine philosophy. Not many make that decision. They go on looking everywhere and nowhere, taking what suits them.

If I said, "Here you can smoke, drink, hold orgies, and still be acceptable", you would think this Teaching utterly marvellous. As I say, "Ah, no! Here you must set aside all your preconceptions and bad habits, here you must make an effort", very few accept, they would rather defend their prejudices and habits. Where will you find someone who is willing in one fell swoop to drop all his stupidities and weaknesses and embrace the divine philosophy of Christ? There are not many.

Le Bonfin, 26 August, 1973

Chapter 16

The Personality Devoured by the Individuality!

"In life all creatures try to absorb other creatures in order to be nourished. Plants live on the ground and feed on the minerals they find there; animals eat plants; men eat animals or at least their products, and who eats man? There are entities who feed on man or, to be exact, on his emanations. In this manner you can go all the way up the ladder to the Lord, who nourishes Himself with celestial Beings."

The fact that creatures eat each other, that all they think about is swallowing whatever they come across... may seem strange to you but it is true. Look at the way animals gobble each other up... and man too has not only little creatures that gnaw at him, but savage beasts, apparent only through a microscope. If a man does not destroy these beasts, he is the one to be destroyed.

Now, it is the same thing for the personality and the individuality. If you give too much importance to the personality, it will swallow the individuality and it all will be over for you. But if you emphasize, if you stress the importance of the individuality, it will little by little eat the personality, which will then grow weaker and weaker until the individuality will be able to manifest itself freely. Wherever you look in the world you see war and carnage... if good is not strong enough to swallow evil and digest it, evil will take over.

This question has never been properly understood. In general people say, "I must have the upper hand, if not, others will get the best of me...." That is true for the lower planes, indeed it is the law of the jungle. But because there are creatures who devour each other in the jungle, in the ocean and in the swamps, it does not mean that man should do the same! We think it normal to fight each other continually because certain thinkers have described the state of mankind as "dog eat dog", but this is on the lower planes. The higher one climbs into the higher spheres the more you see demonstrations of love and self-abnegation, with sacrifice taking the place of the egoism, hatred and cruelty that reign on earth. The world is a theatre of strife and war, but the sun is the symbol of Heaven where there is nothing but love, light and peace. Those who say that the universe is governed by the law of the jungle are right, but only as far as the lower regions are concerned: thus they are only fifty percent right!

Take a child in a family. To begin with, all he can do is eat, cry, demand... he has no thought for anyone but himself. That is what makes him subject to the law of the earth, which is to take. Once he becomes an adult with children of his own, he enters a different system and obeys another law: the law of giving, of sacrifice, the law of love. Men follow formulas which are not true and which do great damage as they spread. People adopt them without realizing that another nature exists which is generous and kind, and so this better nature is neglected, man doesn't know enough to cultivate it and develop it, because he is subject to the personality. He repeats with everyone else the formula, "Slain or be slain!", reasoning according to the law of the jungle and the swamps which is to devour each other.

Now, if we want to make an effort and do the work we should be doing, we must restore to first place the divine nature now lying dormant in some corner... then man will become a divinity, the bad in him will be absorbed by the individuality. If you think the individuality is never hungry and never eats, you are wrong. It eats, and with what it eats it produces Light.

There is an example having to do with food. How is it that when a criminal or miscreant, eats food (all food is divine by nature), it reinforces his wickedness and his desire to kill? Why does the food not make him good, or at least better? Because as he eats he transforms the food into his own nature! It all depends on the state you are in when you eat; food becomes whatever the man is. Wicked people are not improved by the food they eat, they become even worse, whereas good people become better; each transforms the elements into his own nature. That is why the Initiates pray, "Lord God, I consecrate myself to Thee, all I want is to work for Thee! Take me in holocaust, as a sacrificial victim, and absorb me...." They know that they will not disappear, they will not be annihilated, that God will transform them and they will become like Him, they will have His Nature.

Do not be afraid. When I first told you these things you thought, "How can he talk about God as though He were a cannibal who swallows even the Seraphim? No one has ever talked this way, it's really terrible!" Well, you shouldn't have been in such a hurry, you should have waited until I could explain! God does not destroy His creatures, He is not like Saturn who kills and devours his children. Where does the legend about Saturn swallowing his own children come from? They say it is to remind us that time (the Greek word for Saturn is Chronos, time) destroys all things. Yes, but there is still another interpretation, a deeper one, which is that the Creator has a right to eat His own children, His creations or creatures. A writer has the right to burn his own books and a sculptor has the right to destroy his statues, no one can stop them... but there is something, a meaning there that people have not seized. The Lord destroys his creatures in order to transform them, just as the French government during the war asked the French to donate their knives and forks, spoons and other metal objects to be transformed into guns... for in time of war, guns are more important than forks and spoons!

We have a right to destroy something only if we are able to replace it with something better and more beautiful, otherwise we have no right. I told you in a lecture once, "You have the right to kill someone if you can give him a better body. If you cannot, you have no right to take his life." Yes, these are laws to know about. Look at the Initiate: he absorbs as he breathes quantities of micro-organisms in the atmosphere; he inhales, he exhales, and as he exhales he sends what he has inhaled back out in the form of light, love, health. How simple it is for me, how clear, how obvious! But when I open my mouth to explain it, I realize that you are used to seeing things completely differently and the words appear ridiculous, monstrous.

Take another example: our blood. There is a constant fight between the red and white corpuscles to protect the human system against intruders, bacillae, viruses of all kinds. Should the intruder get the upper hand, the man falls ill; when the defenders win, he is in good health. This war goes on ceaselessly in the whole organism without man's knowledge, he is not conscious of these extraordinary exterminations. But one thing he is unfortunately even less conscious of is that when he chooses a chaotic, disorderly, excessive, senseless way of living, he is the one who releases the forces of disintegration that destroy him, that tear him apart and make him ugly. That is why he must change his philosophy, he must embrace the light, he must decide to work according to the divine laws... whereupon he will strengthen the guardians of his system, and the destructive forces will be neutralized. We ourselves determine the factors for good and bad by our way of living. It is therefore up to us to decide which side we reinforce.

In the past I have already given you some well known examples. Take the case of someone who has tuberculosis : if he surrounds himself with good conditions, if he behaves wisely and is careful not to overdo things, he will thereby make it possible for his system to defend itself against the bacillae and neutralize them. His state will improve, he will grow stronger, he will get well. But actually, he will never be completely healed, the bacillae are dormant, chloroformed, and one day, especially if the convalescent begins to live his old disordered life again, they may wake up and his system will not be able to fight against the infection that again ravages his lungs.

What I am saying here has been verified many times, only no one has drawn the proper conclusion : it is we ourselves, for as long as we live, who are the essential factor. The medical world is concerned only with the physical plane, with prescriptions for medicine and diet or rest as the case may be, or with removing organs or grafting onto other organs, but it does not take into account the role played by our thoughts and feelings. That is what makes the good or bad working of the organism. Medicine has not gone that far yet, but one day it will, it is moving in that direction. To know this fact is absolutely essential.

Le Bonfin, 27 Avril, 1973

Chapter 17

Call On Your Allies...

"We ourselves are not well-armed, our weapons are help-less in the fight against evil. Evil, on the contrary, is very well-armed, it has a whole arsenal to fight us with, before which we are powerless. That is why, confronted by the unevenness of the battle, we should ask God to become our ally. What do countries do in time of war? They seek allies, and this immediately renders them stronger than the enemy. It is instinctive with men, this ancient custom of asking associates for help, they know they are too vulnerable alone. We must realize that evil has too many resources, too much knowledge and power, we cannot possibly conquer it alone."

I would like to add a few words today about this instinct that makes us seek allies. It is so true. Even in families, when the mother wants something her husband does not want, she tries to convince the children to be her allies and join her against him... and the father does the same. Everyone knows instinctively the advantage of having allies. In politics, in business, in offices, that is all they do!

We need an ally who is stronger, richer and more powerful than we to join us in our effort to conquer the personality, and the perfect ally is Heaven itself, always near by. You say to the Beings above, "Look, please come and help me because I am so weak, I can't seem to do it alone." The higher up you are the more chance you have of conquering those below. Look at planes, how much more rapid, how much freer they are, with better vision and infinitely better possibilities for destroying a city or country in a single attack than the army beneath them. The military have always realized how important it is strategically to face the enemy from a higher position. Why can't we see that it is the same for the inner life?

The personality is linked with the earth, the depths, the spirits underground, and is therefore lower down than you; the individuality, linked with Heaven and the heavenly Beings, is above you. You who are between the two, should lean towards the higher Beings who have much more power than you, and ask them to help you. Without the support and collaboration of the higher Beings you will always be badly treated, tormented and misled by the personality, for it is very strong, it holds on! For that reason I am always urging you, "Ask Heaven to come and help you, link yourself to the Beings above, they will protect and help you!" You become a spectator watching the battle between Heaven and the personality within you! You will see that in the end light, peace and order are installed and the personality is subdued and obedient, at last in its rightful place... not so much because of your efforts, but because others who are better equipped came and helped you.

If you do not work this way (by far the most effective way), and if you go on trying to overcome the personality by yourself, you will be exhausted uselessly, for the personality is indefatigable and enormously crafty. Even if you dominate it for an instant, right away it comes back into the fray from another direction! We see this in history, when one country succeeds in conquering another one. You think that is the end but it is not, the conquered peoples go to work clandestinely to organize the resistance, and, as soon as the occupants relax for a minute, they are attacked. And this is what happens in the psychic life : you think you have overcome the personality and become its master, but not for long, your victory will be short-lived. You must be vigilant all the time ; the personality is only waiting for the right moment to reverse the situation.

How does the personality handle the problem of sexual energy? You say, "Oh, that's all over, I will never kiss another girl!" and you believe it! But the personality plays a trick or two on you and there you are... in the same old difficulty. You went somewhere to have a drink (to celebrate celibacy), one thing led to another, and so on. I repeat, the personality is very, very clever, very wily, very intelligent, it knows how to move in at the precise moment when we least expect it. So, instead of fighting it continually and spending all our time thinking about it, it would be better to concentrate on something divine, a divine image, because in that way you attract the forces that will help you to overcome it. This is an extremely powerful and effective method I am giving you today : to concentrate on what is positive, powerful, perfect, instead of thinking all the time about the personality.

I say to someone, "Look how unwise you are! Here you are thinking about somebody who has injured you in some way, and so you are criticizing him, enumerating his faults... you never stop thinking about him! You carry his image in your head all the time, like an icon! Do you know what this does, what effect it has on you? By thinking about this inner image all the time, you attract the very faults you are criticizing, the vices and deficiencies of that person... until you become exactly like him! That is the danger of thinking about your enemy. Since you find him so unpleasant and repulsive, why think about him at all? But you take him everywhere you go, you try to make the whole world detest him also, without realizing what this is doing to you, what a bad effect it will have on you. Forget this person! Choose an image which is beautiful, splendid and perfect, and then go to work to be-

come like it, beautiful, splendid, full of extraordinary qualities. You will be radiant!

We think we can conquer an enemy by talking against
him. No, to conquer him we must forget him and having forgotten him, create another image to take his place. In that
way you become radiant, intelligent, strong and powerful,
and that is what will disarm him. You cannot conquer an enemy by carrying his image around with you because he will be
the one to win, sooner or later. Nor can you overcome wickedness by being wicked, cruelty by being cruel, jealousy by
being jealous, anger by losing your temper, and lies with other
lies, for then you identify yourself with those things, you
come down to that level and belong in that category.

It is the same for the personality and the individuality. If
you busy yourself with the personality, you become like the
personality, you are too close to it. It is a law that we grow to
resemble what we look at all the time. The Initiates have always known this law and based all their actions on it. They
say, "Contemplate only what is beautiful, luminous, perfect,
and one day you will be like what you contemplate!" That is
the whole reason behind contemplation. If we are forever
thinking about the personality, we grow the same claws,
hooves, fangs, thorns and darts! Yes, the personality has terrible weapons, one bite leaves a permanent scar, one word is
enough to poison you and make you ill.

The best way to stop thinking about the personality is to
form the habit of concentrating on a sublime subject, to link
yourself with what is divine. In that way you attract heavenly
Beings, Angels and Archangels, who have the best arms with
which to fight for you. Then you will be at peace forever. If
what you want is to take control of the personality in a definite way, you must do this exercise every day, several times a
day. Then it will last.

And yet we should not kill the personality; it can be an
excellent servant, it works remarkably well. Only, when it is

the master of the situation, it leads one into error, because all it knows is how to crush others, to trample on them and despise them. A cross word from someone makes it advise you, "Teach him a lesson, don't let him get away with that, hit him in the nose!", and you rush to obey him. The individuality gives other advice: "Don't get upset, old boy, that's the way it is, but you know what to do, you know how to use this poison and transform it into gold... you are an alchemist!" And there you are, undertaking something wonderful! The individuality continues, "Why weep for hours on end when you have this excellent chance to do a most important work on yourself? You should thank God for this person was surely sent to give you the opportunity to grow. And you're weeping? You are too stupid! This is the time to shake yourself and do something, not to remain feeble, vulnerable, insignificant all your life."

We think we are strong, but the slightest opposition lays us low. How often I see people collapse when they thought they were invulnerable! I say to them, "If you are so strong, so invincible, why are you in such a state? Recognize for once how weak you are, and look for allies... not here, look for them above!" People never think of looking above for allies, they always look on the lower planes, in the family, in society, in nightclubs! No, you must consult those who are competent. When you want to carry off a big robbery, you look for people who know how to open safes! You look until you find them, and there you are with a valuable associate! Why not know the right person to turn to for help in your fight against the personality?

Dear brothers and sisters, you must have this wonderful ambition from now on, of attracting the celestial entities so that you can become conductors of light. That is the only real ambition, the only legitimate, praiseworthy ambition! People say of someone, "That's an ambitious man", and he is immediately classified: it means he seeks power, money, a place in society. I am not speaking of that kind of ambition. Only one kind is valid, and that is the ambition to become like the Lord. Is that ambition, should we call it ambition? I use the word to make myself understood, but the word ambition is always slightly derogatory; to satisfy one's ambition means you are not afraid of upsetting things, even if you have to step on corpses! No, I am not using the word in that sense. I mean it in the sense of a high ideal... never to be content with what is easy to obtain, but always to aim very high, very far....

The methods I give you are to be taken seriously, for there is no more pressing problem than the personality and the individuality. No problem concerns us more, we meet it all the time, we face it countless times in our daily lives. Nothing, nobody, no job, no event, not even our family and friends, are as close to us, as permanently stuck to us, as the personality and the individuality. We live with both of them without cease, so it is worth taking the trouble to study them! I will convince you with my arguments, I will make you understand these irrefutable, essential Truths. At the moment you are with them night and day but unconsciously, as though they didn't exist, you have to think of so many things that are more important. But soon you will understand that it is extremely important to have a very clear idea in your mind about the personality and the individuality, for you will have to spend your life confronting them.

Le Bonfin, 29 April, 1973

Chapter 18

The Further Down, The Less Space

"There are no limits to the heights we can reach when we go upward. Good is unlimited in time and space, but evil is limited, both in space and in time. God has not given evil the right to last forever. When we go toward the positive pole, we come into the boundlessness of Infinity and Eternity, Immense Space, God. God alone is unlimited, all the rest is limited."

According to this idea, one could say that death does not exist. You say, "But that's not true! Look at plants, animals, men... they all die, even objects deteriorate." Yes, but life does not disappear, life is always there. Why is it that death is *not* always there and that life exists in spite of death?

Death does not exist, there is nothing but life, life has existed since the beginning of time. Death is but a variation, an aspect of life. Death is a change of place, a voyage, a transformation, one energy transformed into another energy. Why follow an erroneous philosophy, a way of thinking that puts death and evil in first place, ahead of everything else, even life? Of course, if you identify yourself with matter, the physical body, the idea of death will be of utmost importance. The body being vulnerable weakens, falls ill and dies, and, if you identify with it, you too become progressively weaker until in the end you disappear. But if you identify with the Spirit, the Spirit being indestructible, immortal, luminous and powerful, you too become invulnerable! That is the advantage of adopting the philosophy of the Spirit.

Everything depends on how you look at things, which is something man has not yet understood. He has been brought up on ideas that weaken him, that make him stagnant and annihilate him; this is what we call education! The old ideas must now be replaced by new ideas, those that bring man power, strength, elevation, and bring him closer and closer to God. Never identify yourself with the physical body, with earth and dust, with things that are vulnerable, that grow old and die. Identify with what is holy, and above everything else, with the divine spark: the Spirit. That is Christ's philosophy. The Christ said, "My Father and I are one". Christians have never understood that these words contain the reason for identification. People identify with the physical body which is dust and which will disappear sooner or later. Night and day they repeat this mistaken identification process, growing weaker and weaker, they demolish themselves. If humans are weak, which they are, it is because of the materialistic philosophy they believe in. They say, "Ah, if I had this or that, if I had money, a car, a garden, I would be better off...." No, whatever he may possess, as long as man has the same philosophy, he will be no more than dust. What counts is the Spirit, the Spirit's understanding of all things.

That is what we are bringing to the world, the philosophy of the Spirit, the best philosophy. Those who understand it will make every effort to adopt it and put into practice the things it teaches... and be transformed! Of course, most people would prefer to make money, they think that will bring them everything, but no matter how much they have they will still be buried, just like the others! The man who adopts the philosophy of the Spirit will become immortal, he knows that his body is not his real self but rather a piece of clothing, used clothing! What happens to you if you identify with your clothing? Would you dream of identifying with your car? You do not say, "I am this car", for you are not the car, it is merely a convenience, a means of transportation. Do you identify with your horse, or with the rider?

"My Father and I are one", means that Jesus identified with his Father. "My Father" is not simply any father, it is our Heavenly Father, eternal, immortal, all-powerful. We must now imitate Jesus and identify ourselves with our Heavenly Father, the Spirit.

The personality commands formidable forces which you must learn to use exactly as man learned to use the natural forces of water, wind, electricity, etc.... When men didn't know how to use them they were constantly being victimized; now they do all kinds of miraculous things with them. Would it not be the same for man's inner forces? Once the disciple knows how to take the forces that frighten him and threaten him, he will become a resourceful engineer, inside him there will always be running water, lamps, windmills and everything else he needs! In an Initiatic School, instead of fighting against what disturbs him and trying to eliminate it, the disciple makes use of it! Christians are always talking about extirpating, annihilating evil, but the battle is too uneven, they lose their strength, their best assets, and are then condemned to permanent weakness and insignificance. That is not the best education. We must replace it!

You see how many changes we need to bring about, how many things we must learn! To learn how to use vanity, for instance, or sexual energy, or anger. My vanity does all the work for me! If I were not vain, I would never do a thing, but long ago I harnessed my vanity and put it to work, and now it does wonderful things for me... why uproot it? On the contrary, I caress it, give it a few pats now and then, a little food, and put it to work. It moves mountains! I have never said I had no vanity, I am proud of having a vanity that does so much and it is the same with other forces also. Those who fight to uproot their faults will never do anything, there will be no results. My advice is, instead of fighting against your vices, put them to work! Now, do you see how different our methods are, methods that bring results!

By fighting against sexual energy as most religions advise us to do, and repressing these energies mercilessly, you end up as a case for psychiatrists and psychoanalysts! But if you know how to use it, how to put it to work, you may become a highly exceptional being, a genius. Leonardo da Vinci, for instance, knew how to sublimate the sexual energy. He never married, he had no adventures apparently, but his paintings reveal great sensitivity to feminine beauty. One day, a young count brought his ravishingly beautiful wife to meet Leonardo and asked him to paint her portrait. The artist looked at her and exclaimed, "How beautiful she is! A great beauty! What would I not give to paint her in the nude!" The husband and wife exchanged looks, the husband indicated that she could do what the artist asked... and the artist went to work!

Leonardo da Vinci was not repressed as were the ascetics, the hermits, whose idea of woman was that she was the incarnation of the devil, to be avoided at all costs. History does not say whether or not he knew the secret of sexual sublimation, but I looked into the question and found that the tremendous discoveries he made were due to the fact that he sent this energy up to his brain (sublimation) and, because of that, was able to give wonderful discoveries to mankind.

On the subject of love also, we bring mankind a new instruction which it badly needs. The old instruction, the old puritanical pedagogy, filled hospitals more than anything else! We must become like all true artists: they look, they appreciate, and they sublimate their energy and render what they see divine. The others, poor things, merely succumb! The true artists are the Initiates.

I will give you another precise, scientific, irrefutable argument to show you that the individuality represents infinity and freedom of movement, whereas the personality is limiting, it paralyzes us in increasing degrees. Dante, in his *Divine Comedy,* depicts Hell as a reversed cone, as space growing more and more narrow. The personality is also a cone, the further down you go the more limited, crushed, compressed you become... until you can no longer move at all.

The four elements of matter, fire, air, water, earth, present the same idea in a different way. The further down you go, the more dense and dark matter becomes, and the more difficult it is to move. A mole on the ground, a fish in the water, a bird in the air, all move, but not with the speed of light, which takes only eight minutes and a few seconds to go from the sun to the earth! Nothing can equal the speed of light except thought. Is this clear now? The lower down we go, the more limited we become as to our movements, and the higher we go into the realm of the soul and spirit where matter is more subtle, the more possibilities we have. When you climb, you work with the individuality, and you can go very far, you can expand all the way to infinity, you are free! But when you work with the personality, you come to a stop very soon, there are too many impediments.

I am not inventing this, I am merely deciphering Nature, for Nature speaks! You must learn to read the Great Living Book of Nature. I say to those who have degrees and diplomas, the professors and pundits, "You have not yet begun". "What, not begun?" "No, for you can neither read nor write." "What are you talking about?" "The things you read have no great value, what you should do is learn to read from the Living Book of Nature. Can you read it? Or write divine things on the souls and hearts and spirits of your fellow beings?" Take a man : with his "pen" he writes on his wife's "book", and a "child" is the result. A child who is born sickly or violent is the proof that this man did not know how to write!

To the same extent that in the past it was difficult to convince parents to send their children to school, so it will be difficult now to make humans go to a Divine School and learn how to read and write. Everyone all over the world goes to school now, even savages, but it was difficult to persuade them! Well, now it will be the same thing to make mankind accept to go to a new School, a Divine School, and learn how to read the Great Living Book of Nature! Did you ever think of that?

Le Bonfin, 26 July, 1973

Chapter 19

Your Inner Animals

"If you learn to use your thoughts and feelings in a way that reaches your subconscious, then you will know at once when your wild animals start to threaten your domesticated animals, or the forces you work with. A man who has tamed all his inner animals can rely on the domesticated ones to work for him, and he will live in abundance forever."

This thought will seem odd to you, especially to those who do not know that what exists outside of man also exists inside him. A human being is the synthesis of everything that exists, minerals, plants, animals, human beings, Angels. So it is not surprising that there should be animals in his subconscious in the form of instincts, impulses, tendencies. Our instinctive and emotional life is the result of the extent to which we have been able to tame our inner animals and put them to work, just as in the past horses, cows, dogs, goats, cats, lambs, camels, and elephants were tamed and made into servants.

Man cannot tame his wild beasts because of his guilt, he cannot forgive himself for committing the fault which caused him to be chased out of Paradise. The first man and woman lived in Paradise in peace and harmony with the animals; the animals were their brothers and never harmed them. It was Adam who took care of them, they obeyed him and they understood him. Of course, you will not find this in history books, but if you know how to avail yourself of the Akasha Chronica, the Annals of Mankind, you will see that before the Fall, when man was primitive and still had light, wisdom, beauty and power, he was in harmony with the forces of Nature and they obeyed him. Later, when he decided to listen to other voices and follow other influences, he lost his light as well as his power over animals, and a real scission took place, with some animals continuing to serve man and remain faithful to him, and others declaring war on him because they could not forgive him. It was the Fall of man that made beasts become wild and dangerous.

Of course, human beings will not accept this idea, they see no relation between themselves and animals. But I tell you that many of our thoughts and feelings live inside us in the form of tigers, wild boar, crocodiles, leopards, cobras, scorpions, octopuses and so on, whereas others are in the form of birds, meek and harmless. According to the Initiatic Science, a whole fauna lives inside us; if you have been thinking that prehistoric animals such as the dinosaur, ichthyosaurus, pterodactyl, diplodocus and mammoth have entirely disappeared, you are mistaken: they are inside man! You say, "But there's no room!" Yes, because they are in another form in our astral and lower mental bodies. You must understand that we are not talking here about the outer appearance of the animal but his characteristics, his quintessence, what he is known for. Independently of the physical aspect, each animal is known for a particular characteristic: a rabbit's way of eating, his long ears, are not so characteristic as the fact that he is timorous, a wolf's thick neck or his ability to run great distances are not as characteristic as its destructiveness, it finds its way into the sheepfold, not to kill one lamb in order to satisfy its hunger, but to strangle as many as possible.

A lion is best known for pride and courage, a tiger for cruelty, an eagle for its piercing sight and its love of heights, a goat for sensuality, pigs for filth, dogs for fidelity, lambs for gentleness, cats for independence and suppleness, cattle for patience, camels for sobriety and roosters for combativity... we could go on and on.

It is therefore in the form of their qualities or deficiencies that we have animals within us. Furthermore, people's faces often resemble animals. The great Swiss physiognomist, Lavater, showed clearly in his drawings how much humans resemble pigs, goats, monkeys, dogs, horses, camels, hens, fish, etc....

If you observe yourself, you will find quantities of animals inside you; one feeling is a lion, another is a scorpion; our thoughts are like birds, our emotions like reptiles and quadrupeds. The vast realm of emotions also includes elementals, larvae, disembodied spirits. Nor have any of the peoples and tribes that have existed disappeared, they are in human beings, in forms that are very difficult for you to understand or envisage. For the moment, I am merely mentioning it because I want you to be aware that everything in the universe is also in man, mountains, lakes, rivers, oceans, swamps, trees, flowers, crystals, minerals, metals, and, of course, animals.

Now, what is man's role? Man has been given a mission, that of bringing his entire being into harmony, of making his wild animals tame and obedient, willing to work for him. People who have domestic animals working for them on the land transporting heavy loads, become rich thanks to these animals. Yet men do things with animals that are not strictly correct or just, such as cutting them up for food, selling their flesh and their skins....

I have described the conversation I had one day with a writer who come to see me. He asked a question to which I re-

plied that there were invisible spirits in the world who devoured humans. He became highly indignant and denied it,
saying it was unfair and so on.... "And yet," I said to him.
"You are a writer, you should know these things! You know
human beings... how do they behave toward animals? Don't
they eat their meat and sell their skins? Well, is that fair? No.
Well, they are exactly like certain spirits in the invisible
world, who work unceasingly against humans, milking them,
selling their flesh, their fat, their bones, their tusks, their fur
and their young. Even if you refuse this idea of entities in the
astral and mental worlds, you have to admit there are viruses,
germs of all kinds that nourish themselves on man." He was
aghast.

Humans have a tremendous role to play in Creation, but
at the moment they have gone far from the Source and forgotten their role, they no longer know their predestination, and
behave like animals tearing each other apart, devouring each
other. They may appear human on the surface, with their fine
clothes, their medals and decorations, their houses filled with
books and works of art... yes, to show their high degree of culture. But inside they are crawling with all kinds of lower
forms of life, instincts, appetites, cruelty, hostility. Do not
think we are ever rid of wild animals! Because we cannot see
them in their physical form, we think they are no longer
there, but they are... in the form of jealousy, vengeance,
hatred, violence... in our thoughts and feelings. It is our task
to tame them, to teach them to be wise and gentle and willing
to serve. Anger, vanity, sexual energy all must be tamed and
trained... and put to work for the Good.

Look at what happens in primitive, uncivilized areas: if
children, domestic animals, fowl and livestock are not constantly watched and protected, wild beasts seize and devour
them. In the same way, if a man does not know how to preserve himself, wild beasts come out of the forests and attack
him and his children. What children? The good thoughts,

good feelings, good intentions, good influences he has brought into the world. If a man is not protected, hostile forces steal into the barnyard, the pasture, the playground... and then he wonders why he is in trouble. How often I have seen this! Someone says to me, "I had made wonderful plans, everything was going to be so marvellous, but then I lost my inspiration, my enthusiasm...." I want to say, "Because you were not vigilant, because you were asleep when you should have been watching, the wild beasts came and ransacked everything." I refrain from saying it because I know I will not be believed! And yet, how else do you explain this disappearance of our inspiration, our momentum?

The question of the inner animals is an important one. We must be very strong in order to dominate them, with the kind of strength that is the result of love and purity. In India, for instance, yogis meditate all day in the forest with never a worry about the wild beasts around them. Animals are very sensitive, they can sense the aura, the light that shines out from holy beings. Humans have little or no sensitivity, but animals sense things. Therefore, if you want your inner animals to obey, you must increase your light, your purity and your love, you must come closer and closer to God. Then the animals and everyone else will recognize your authority and be obliged to listen. Otherwise you can try everything you like, they will not obey.

I am not the only one to have made this discovery... thousands have discovered that animals obey only if their owner is on the path of light (I am speaking of inner animals, you are not apt to meet savage beasts in your daily life). When I was travelling in India, I was warned not to walk anywhere near where there were tigers. I did walk in those regions regardless, and the funny thing is, I never saw a single one! Why is this? Either they were afraid of me because they sensed my cruelty and ran off, or I did not merit to see any! Therefore I have no idea whether I would be able to tame a wild beast or not.

During the era when they persecuted Christians, wild beasts in the arena sometimes spared their victim, whilst others would be immediately torn apart... not necessarily because those victims were impure or had abandoned their faith. Death comes when it is your destiny to die, as the result of snakebite, the capsizing of a building, an accident, boiling water or poison, drowning, a bullet, a knife wound... all that is determined ahead of time and for a specific reason. Each person is linked with one of the four elements in particular, and, depending on which one, death will come from the earth, water, air, or fire.

To be able to domesticate one's animals is a worthy task, success brings great advantages. Once you tame the inner animals you can then go on to work on exterior animals, but you will have no effect on them until you have succeeded within yourself. I have watched animal trainers the world over, and the fact that they know how to dominate wild beasts does not always mean they have succeeded in controlling their inner beasts. When they use fear to discipline animals, the animals obey because they have no choice, but the minute the trainer relaxes his control, immediately he is pounced on. When I was a student in Varna, in Bulgaria, a snake charmer came to demonstrate before us one day. He was dressed all in yellow and carried bags filled with snakes of all kinds, including venomous ones. He let some of them out onto a platform and fixed them with the most intense stare. His look was extraordinary, the snakes shrank from it. We were duly impressed, but a short time later we heard that he had been bitten to death by one of them. He must have relaxed his vigilance. If he had controlled himself, and above all, if he had radiated the love which makes even the most ferocious animal yield, he would not have been bitten.

Well, let us leave all that now. Hold on to the thought that we, as human beings, have within us all the reigns of Nature:

the mineral kingdom corresponds to our bone structure, the animal kingdom to our circulation, human beings to the nervous system. Above is the auric system which is still more subtle than the nervous system, and represents the limit between the human world and the world of Angels.

All Initiatic Schools have agreed on one point, that man is a summary of Creation, he is the microcosm, the little world in relation to the macrocosm, the large world or universe. This fact explains the work of the Masters. Since all the universe is inside them, they can touch Heaven itself! But if Heaven is within man (and it is), unfortunately Hell is also. Alas, man harbours all the devils of Hell! Fortunately they are somewhat hidden, paralyzed, chloroformed, practically motionless... but if something revives them, as a snake is revived by heat for instance, then instantly they bite. To make a snake harmless, it must be chilled; heat makes a snake dangerous. Why are men who are especially "hot" the ones to be bitten? By "hot", you know what I mean. I am not referring to friendly or brotherly warmth.

There exist several different kinds of warmth, of which one in particular arouses the serpent: sexual energy. How often are we bitten from having overheated the serpent! Initiates prefer to keep it cold so that it will be harmless. That is what cold is for! Be a little cold where sex is concerned, but keep warmth in the heart. How do you warm the snake? You don't need me to tell you that! People know how, with alcohol, aphrodisiacs, certain attitudes, the right word or look, perfume, music. Then the serpent comes to life, and the first thing it does is to bite that person. That is why people say, "Ah! You are bitten!" Bitten by what? By the inner serpent, the dragon if you like.

Do you remember the lecture entitled, "How to Measure Oneself Against the Dragon"? The dragon is in us, and in direct opposition to the dragon is the dove. The meaning behind the symbol of the dove is directly opposed to the serpent.

That is why the dove and the serpent hate each other; the snake hates the dove and the dove is afraid of the snake. The four Holy Animals, the Lion, the Bull, the Eagle and Man, correspond to four signs of the Zodiac, Leo, Taurus, Scorpio and Acquarius. Why does the Eagle correspond with Scorpio? Because in the past the Eagle was in that place but since the Fall of Man the Eagle has been replaced by the Scorpion, the symbol of the fallen Eagle. The Scorpion must become at once Eagle and Dove... and there you have the process of sublimation!

Le Bonfin, 14th August, 1973

Chapter 20

But Which Nature?

My worries are increasing, dear brothers and sisters! You are becoming so numerous, there are now so many of you, that the Rocher has become too small! We must enlarge it. If each one of us brings a rock or stone and a little cement in his pocket when he comes to the sunrise we will soon have enough to make a large smooth platform for us all to sit on. You see, together we will solve the problem!

In any case, this morning the sun came out for a few minutes... we cannot complain.

Now, as today is Easter, I might say, "Christ is risen!", but will you be convinced? Perhaps in two or three hours when you are comfortably installed in the Hall you will believe it, but how can you believe in a risen Christ when you are cold and uncomfortable? The Russians have a saying, *"Dengui iest, Christos vozkresse; deneg niet, smertou smert!"* which means, "Where there is money, Christ is risen; where there is no money, Christ is dead!" A wise proverb! Ah, the Russians know how to tell whether Christ is risen or not: money! Bulgarians greet each other on Easter Day with the words, *"Christos veuzkressi"* or "Christ is risen!", to which one answers, "Na istina, veuzkressi!" or "Risen indeed." But like all traditions this one is rapidly being lost because of the materialistic culture we live in.

Isn't it marvellous to be so many, all together like this? When we become even more numerous, you will see how alive we will become! A tremendous awakening for all of us, simply because we are here together! Of course, there may be a few inconveniences as a result of being in a crowd; collective life is wonderful in many ways, but one must be very vigilant, alert, and enlightened by a High Ideal. Then there is no danger, we will be protected, upheld and encouraged as we rise toward the spiritual heights. That is the Initiate's Ideal. I know many of you have come with tendencies that you inherited and which are not entirely orthodox, perhaps even with an eye to taking advantage of someone here and satisfying your own interests. But I continue to hope, I count on the power of light to change all that.

We all bring with us the old tendencies we have inherited from the past, they are permanently installed in us... no one is free from the past. The difference between people is that those who are more evolved control their lower tendencies and *want* to work for the highest Ideal, whereas those who have no light, no impetus toward a High Ideal, continue to manifest their lower passions... this is only normal. Without the light, how can you expect them to do anything else? What do they have to keep them from following their desires, their selfish, self-centred habits, grabbing what they want, thinking only of themselves and wanting to swallow everything they see? It is normal and natural. In fact those people consider Initiation as abnormal and unnatural! But the divine world does not think that... the Higher Beings consider it perfectly natural.

On this subject men are singularly unenlightened: what is "natural" and what is not? Everyone talks about behaving according to the "natural" laws, following their own "nature"... which is all very well, but *which* nature? For there are two. Many obey "nature", as they say, when they are really allowing themselves to be governed by their *lower* nature and doing everything they can to obliterate the higher nature; others who wish to liberate the higher Self in them do their best to deny the wishes and desires of their animal nature. But confusion reigns in the human mind and that is why I would like to throw light on this subject, a new light, so that men will know of the existence of another nature which is divine and glorious, and which manifests itself in an opposite way to what they call "human" nature, which is in fact the lower nature.

Animals are totally subject to their lower nature, they remain faithful to its laws because animals are governed by their instincts and not, as humans are, by intelligence, conscience, or willpower. Humans have something that makes them different from animals: Cosmic Intelligence has given them the means to oppose the lower nature, to extract themselves from its grip and place themselves under the guidance of Divine Law. It is true that our Teaching is "unnatural" in that it is bent on keeping us from manifesting the egotistic, cruel and unscrupulous nature, inviting us, on the contrary, to conquer it, control it and make it obey, in order to free the beautiful, noble, generous, divine nature.

Take fear: Nature made animals fearful on purpose, so that they would protect themselves from danger. All creatures began by being fearful, and later, when they became more evolved, Cosmic Intelligence intervened and freed them from this impediment, replacing fear with intelligence. It is better to know, to understand, to see things clearly than to remain ignorant through fear. But with animals, their fear saves them from danger, since they do not have intelligence. Man, who has this new element, the factor denoting progress, or intelligence, must not remain fearful as animals, for it is not "natural", it will keep him from evolving. We can then conclude that what Nature advocates and approves *for a certain time* is not designed to last forever. Later on Nature will advocate something else. This is true of many things in life: we use all our strength to obtain something, and then we must strive with all our strength to get rid of it! Wisdom lies in knowing when to hold on and when to let go.

Here is another example taken this time from the religious field. In the Old Testament, it is written that man should fear the Lord because fear is the beginning of wisdom. Then Jesus came to replace fear with love, because He knew that fear would keep man from growing and evolving. The Lord was no longer to be feared but to be loved as a Father. It is the same continual process: the lower nature works in a certain way for a certain time to make humans go in a certain direction, and then the higher nature comes along and sweeps them off in another direction! Is this clear now? This Teaching may be unnatural and abnormal, if you like, but it is better!

Another example: a boy who is attracted to a girl feels a desire, an impulse, to pounce on her... this is natural, is it not? Yes, but what if he goes on forever obeying that nature, what will become of him? He will remain an animal. The moment comes when the other nature intervenes, it whispers, "You must control yourself, you must dominate those impulses now". You could say this is an "unnatural" nature speaking, but the advice it gives is preferable! Or, suppose somebody feels the need for something belonging to his neighbour: his lower nature will urge him to go and take it... he needs it and that's all there is to it! No need for scruples! But if the higher nature happens by at that moment, it will advise the contrary, "Ah, no, no! That belongs to someone else, you must not take it, you have no right! If you do, you will have to pay very dearly." That is intelligence, justice, morality.

Do you see now what kind of place you have landed in? An abnormal school! Now make the best of it... you should have been more careful where you put your feet! I sometimes say to those who come here for the first time, "You don't know it, but you have landed in a place of perdition!" They look at me in alarm and I continue, "Yes, a place of perdition in which you lose all your faults, your fears, your blindness!" Thus reassured, they smile at me.

So there you are: the "natural" nature and the "unnatural" nature. When someone says you have a "little nature", it means you have a timid disposition, you are weak and vulnerable, tearful, whining. The other nature has a strong and cheerful disposition, it says to you, "Here, climb on my back, I'll carry you!" Is that clear? Men all follow their "nature", but the question they must be able to answer is *which* nature, the animal nature or the divine nature? Unfortunately most people are loyal, faithful and true to their animal nature! Yes, firmly convinced that there is no other way, and if you should try to awaken their other nature.... Oh my! Life becomes complicated indeed! But it must be done, the house built by our ancestors in past centuries was marvellous, magnificent for that epoch, but now the time has come to tear it down and build a new house, the old one is mouldy, near to collapse... it must be torn down and replaced by a better one.

Now you are forewarned, you can no longer object, no longer complain, "But what is he asking us to do? Where is he taking us? It will be the death of us!" Yes, death in exchange for life, to die in order to live, as a seed dies in the ground in order to grow and blossom! If it does not die, that is, if it does not refuse to stagnate, to live uselessly, which is nothing but another form of death, it will not live, that is, it will not bear fruit. We too, if we remain behind with the old concepts, the old ways, will never really live. To be really alive, we must die to the old forms and adopt new ones, magnificent ones. You cannot actually believe the Christ wanted men to die? No. "Unless ye die..." means "Unless you change your form, your habits, your way of thinking...", but the Christ who said, "I am the Resurrection and the Life..." could not want us to die! What he wanted was for us to be alive in the way He is alive!

So, whatever we do, whatever we say or think, there is only one thing to do : accept to die to everything that belongs to the lower nature, and be reborn to the higher nature which is divine.

As obedient and docile as we were regarding the laws of the lower nature, so now we must become obedient and submissive to the Divine Laws of the higher nature, the individuality. Of course the lower nature will shout in protest, it will insist on its "rights", it will revolt and hurl threats, because this nature is terrible when roused. You *must* declare war on it and make it submit. And beware... if you are not much more powerful and much better equipped, better armed than it is, it will refuse to obey. That is why you must ask the other nature, the divine nature, for help. In life when someone attacks you on the street, if you shout for a policeman, your attacker will take to his heels. In order to reduce the lower nature to silence, you must appeal to the divine nature for help and when it appears in all its glory, the other nature will sense its inferiority and give in. "I can take care of myself", you say. Oh, my! It isn't afraid of you! It knows you, how many times has it fooled you and got the best of you? For it to give in, you must have the help of a very powerful ally, very powerful indeed.

Therefore call on the divine nature, beg and entreat it to come and help you. All you need is its presence, as soon as it appears, the other nature will be struck dumb and retreat without a word, in complete silence. One power can only be conquered by its opposite power, as proved by poison and its antidote, or heat and cold, etc.... Increase the quantity, the potency of one power, and the other will diminish automatically. This law is faithful and true, one of the great Laws.

Le Bonfin, 14th April, 1974

Chapter 21

Sexual Sublimation

"If you should talk to a sensual, primitive person about a more spiritual concept of love, he would say, 'But that's not possible! If we cannot satisfy ourselves sexually, we will die! It is what makes us live.' Yes, that may be what makes the roots live, but above the roots, the flowers die. It all depends on the person, on his degree of evolution."

I have talked so much about love and presented it under so many different aspects, you will never be able to know it all from one little passage taken from a lecture. You will understand no more than before, you must refer to many more lectures in order to have a clear idea. Each lecture discusses one point in particular.

In any case I think the more one explains this question of love from the Initiatic point of view, the less people understand. Why? Because for thousands of years they have had the same ideas, done the same thing, behaved the same way. They do not realize that Nature inspires in her creatures the need for a certain kind of sexual behaviour, but only *for a certain time*, after which she takes away those needs in order to make her creatures discover other, higher, more beautiful and more spiritual ways in the field of sex and in other fields as well.

.man beings are meant to evolve in every way, in all
.ifferent areas, would they not also be meant to evolve
.ere love is concerned? The next degree of love on the ladder of evolution consists in sublimating the sexual energy by
directing it toward the summit, the head, to nourish the brain
and make it capable of doing extraordinary things. As long as
man does not know how to use this energy for the gigantic
task he is here to do, it will be wasted and he will sink into a
state of stupidity, of animality. Everyone knows the direction
the sexual force goes in, but few know it can be oriented in an
entirely different direction, that Cosmic Intelligence has given
man a system of canals and inner instruments which can
carry this energy upward. People do not know this. Everyone
has these canalisations and instruments inside, waiting to be
used, but they are not yet functioning.

People think of sexual energy as a consuming tension
which must be released. And that is what they want, they do
everything to free themselves of this tension without realizing
that they are losing something very precious, they are burning
up a quintessence of inestimable value for the sake of mere
pleasure! A question: when you build a skyscraper with 150
floors, you need tension to bring up the things that are needed
at the top of the building. The inhabitants in the penthouse
need water to drink and for their plants; by suppressing the
tension, you prevent the water from going all the way to the
top. Look how ignorant human beings are: in order to be free
of the discomfort of tension, they cut off the water supply
and... go to seed.

This tension must be used, not wasted; without it there
will be no energy at the top. The cells of the brain, instead of
being awakened and stirred into action to do a gigantic work,
will remain numb, sluggish, dull, chloroformed into inactiv-
ity, able to function only on the lower level. Man must learn
to control and master himself in order to be strong, powerful,
intelligent. He thinks, but why make the effort? The pill is

there! Before the pill he had to think a little, to pause and control himself somewhat, but now there is no need for self-control and that is why he is growing weaker and weaker in every way. Pleasure devours everything, your forces are dispersed and later you are incapable of doing anything at all. Everything has been devoured, burned, ruined.

How to make men and women understand that in God's Plan this energy was meant to be creative, to produce divine creations! No way to make them understand... all they want is enjoyment and pleasure, ease, no effort. But this form of enjoyment must be paid, and paid dearly. Whereas, if one makes the effort to control oneself, not only is one the richer for it, but it brings a most extraordinary pleasure! No, the word pleasure does not apply here, for pleasure is connected with lower instincts... the words joy, rapture, ecstasy, are better in this connection. Pleasure is nothing glorious, one can be ashamed of it, whereas joy, rapture, ecstasy, are the result of something divine.

Young people have no idea such experiences exist, nor that they are far more enriching than their sordid adventures which do nothing but rob them of their freshness and charm, their beauty, their light. I say to the youth of the world, "You want to experience physical love? Very well, do so. But know that in a little while all the sensations, experiences and experiments will leave you with nothing but regret, obscurity, depression... you will be in ruins." That is what they must be told. And adults also let themselves go, they too slide downhill without realizing what will happen. They could at least try to control themselves. Even if there are no results at first, there will be later, and then they can be proud of having overcome their lower self: a pride that makes one strong and sure.

A few days ago, I received the visit of a man who is fairly ripe in age, and he admitted to me that he was extremely weak, he had no control where sex was concerned, and he asked my advice. I said to him, "I am willing to suggest something to you, but it will be hard for you because you have never used control. You should have tried to train yourself years ago. These exercises are for control: go to the beach, look at the pretty girls; obviously, something will stir in you... it is only natural that it should. But, as you will not be able to satisfy your desire then and there (there are too many people, you don't know the girls), you will be *forced* to control yourself, to use your willpower. If you repeat this exercise several times with success, you need not go on with it; instead try leafing through magazines with pictures of nudes. Again something in you will begin to stir, but this time you grab hold of this feeling as it wakens, and send it up all the way to Heaven, to the Divine Mother which is Nature. If you do this exercise over a long period of time, the day will come when you no longer need to have physical relations. That is a victory, a real triumph! But it takes homeopathic doses over a long period of time, diluted to an infinitesimal fraction of the full dose, but without losing any of its efficacity! You can dilute love to the point where you no longer need physical contact. Then it is spiritual love." Of course, the man was astounded, amazed, and he left with high hopes, but I do not know whether he succeeded or not.

I could say many things on the subject of sublimation of the sexual force, but you would not understand. Be content to take what I have told you, and take it seriously. Whatever people say, there is always human nature to contend with, but people are hypocritical enough to pretend it does not concern them! When others love and kiss each other they make fun and criticize them, but if the truth were known! What do those critics do in secret? People ought to be more sincere, more honest... like me. I see the whole question of sex differently and I say there are ways of dealing with this phenomenon, it all depends on your degree of evolution.

Everything in life is beautiful and full of meaning as long

as you have the right attitude about it, do it the best way possible, and never have to reproach yourself, as people do who behave like animals. How can they be proud under those conditions? They are weak, they are vulnerable, they give in to their instincts, but afterwards they see how unaesthetic it was, and are disgusted. If you can hear what I am saying, and if you begin now to train yourself in another direction as I have suggested, you will discover the beauty, the splendour of the other love, the transparence and luminosity of real love. Night and day you will want to delight in it and fill yourself with it. This love does not last for a few minutes but for all Eternity. You go on loving forever.

With the ordinary love, of course one feels love, but very quickly that love turns to hate, the need for revenge. In the beginning it was gold, but it changed all too quickly into lead, and then nothing is left but ashes, bitterness, arguments, disillusionment. The disciple of the great Universal White Brotherhood must understand that it is to his advantage to experience the higher degrees of love, where he will find freedom. Because of that freedom he will have no need to look for anything from anyone, he will be the one to give ceaselssly to everyone else. It is their needs that make people dependent and unhappy. As long as you need to make exchanges on the lower levels of the physical plane, in the personality, you forfeit your freedom, and you suffer. Of course, not everyone is able to control the sexual energy and experience the higher forms of love. Before throwing yourself into an adventure it would be well to reflect, especially to know yourself. If you feel that you still need physical pleasure, it is better not to abstain too brusquely, otherwise it will be worse. But if you are a little more evolved and feel the need to live something more spiritual, to come closer to understanding the subtleties, the splendour of the divine world, and to help others by your love, then you would do well to choose this path. It is not for everyone, I do not advise everyone to take this path, because I know very well what the results will be. What will happen to a couple, for instance, if one of them should decide to live on a higher plane and love more spiritually, and the one that is left, who cannot get along without physical pleasure, makes it into a tragedy? Of course, I will be held responsible! I know it is dangerous to talk about these things this way, but I do it to enlighten those of you who want to evolve. I must do this, but I am very conscious of the danger I run of being misunderstood, the danger of creating hostility. I risk a great deal, but no matter, I must do it.

All I ask is that the truth of these words be heard by my listeners, so that they realize my aim, which is not to separate families, but to enlarge consciousnesses, the consciousness of all men and women. If love as the world conceives it had brought marvellous results, there would be nothing to say, no need to speak, but look at all the divorces! Look at all the separations, dramas and tragedies! Even when they are together, husband and wife are both thinking of someone else, some film star or rock singer. Both are deceitful, each betrays the other.

Now, say that I have here two bottles: I ask you to imagine one as a boy and the other as a girl. Because they do nothing but draw upon each other, very quickly these two bottles are empty and tossed aside, and each one goes looking for a new bottle. That is the old form of love. To drink from a bottle until it is empty and then throw it away. With the new Teaching, the "bottles" are connected with the divine Source, you can drink and drink to your heart's content without ever draining the bottle, it will always be full because it is constantly being filled at the Source, the divine Spring.

This means that instead of loving a man or woman's personality, their physical body, you love their Spirit, their soul, the Source, God. That love will last forever, even when you are old and grey and wrinkled as old apples you will still love each other, because it was not the physical body you fell in love with, it was the being, the divine reflection of God Himself. In the Teaching of the great Universal White Brotherhood, men and women learn how to love each other. Through the woman he loves, the man seeks the Divine Mother and reaches out toward her in order to receive her energies, her light and joy; through her love for her husband, the woman climbs all the way to her Heavenly Father. Never will this love end. Men and women who content themselves with seeking love on the physical plane should not be surprised when their love does not last. It is natural, is it not? Why love emptiness, someone who is empty inside? And that is what happens when all you love is the physical body, the day comes when you feel the emptiness, the void. But if you love each other mutually for an idea, an Ideal for which you both work, it is plenitude, you will never again be separated. *That* is what we must tell the young.

How many youths come and tell me about the girl they love! I ask the boy, "What is it you love about this girl?" "Oh, her face, her legs, her figure, her eyes...." "Well then, if that's all, believe me, your love will not last." "What? Why?" "Because you haven't taken the trouble to discover what this girl thinks, what her ideas are, her tastes, what she aspires to. All you love is the outside. In no time at all you will be sick of all that, what you find charming and pretty now will vanish into the background and you will discover her faults, they will make you forget her beauty and want to leave her as quickly as possible!" I ask another one the same question: "What do you love about this girl?" "Oh, I like the way she thinks, I like her taste for the spiritual." "How wonderful! And physically?" "Physically, she doesn't appeal too much." "Well, don't worry. When you love someone for his way of thinking, feeling and behaving, when you love his soul and Spirit, you become so attached, you love him so much, that in the end you are together even physically." I have seen people be almost repulsed by each other at first, and little by little, because of their intellectual and spiritual understanding, fall in love physically as well.

That is why I say to young people : before getting married, try to find out if you have the necessary affinities in the sphere of ideas and emotions. The physical plane comes in third place. If you marry simply as a result of being physically attracted, when the other side begins to appear, you will start arguing and fighting until you actually hit each other. This is the way many marriages finish, unfortunately... because the two people were too stupid, or because they were not well advised. Parents often do not know how to advise their children ; instead of enlightening them, they say to their boy or girl, "You're on your own!" I realize young people do not always listen, even to sound advice, but at least one day, after failure and catastrophes have taught them a lesson or two, they will remember your advice and follow it. I know that explanations and advice are not always effective... who knows it better than I? But afterwards, when they have come to the end of their rope, they will remember your explanations and begin to think.

Ah, how many more things to say, to explain, to throw light on! But in summary, I say this : I advise the brothers and sisters of the Universal White Brotherhood not to tie themselves up for life with boys or girls, men or women, who have no spiritual ideal, because their love will not last, or for it to last, they will have to give in to all the desires and whims of their partner, and sacrifice their divine ideal to the physical. Many have done that, even here in the Fraternité. They have fallen in love with people who had no ideal and sacrificed everything the Brotherhood would have done to prepare their soul for the future. Now that it is too late, they suffer, they regret their action. Believe me, dear brothers and sisters, do not tie yourself down to people who have no ideal, no desire for the spiritual life. Otherwise you will suffer.

Le Bonfin, 15 September, 1974

Chapter 22

Toward Universal Brotherhood

I

"During the silence, the disciple should concentrate on participating in the work of the Master. Everything the disciple does in his life should be for the good of mankind, for the Universal White Brotherhood which is above to be realized below on earth. The Brotherhood lightens man's task on earth and at the same time gives him the impetus he needs to reach the glorious heights."

Many things can be added to that thought, dear brothers and sisters. What does it mean, for instance, that during the silence the disciple should participate with his Master in the work? And what is meant by, "Everything the disciple does in his life should be for the good of mankind..."?

One glance at society, the family, our neighbours and ourselves is enough to show quite clearly that everything in life is designed to satisfy our human nature, which is actually nothing more than our primitive animal nature with all its savage, prehistoric instincts and desires. All our rules and standards, all the goals of our society and education were conceived with the idea of helping man to achieve material goals, to triumph over others, to profit and get ahead, to grab everything for himself and never mind the others! Our rivalries, quarrels, violence and hatred that grow into giant wars are the result of everyone thinking, "Me, me! All for me! Grab what I want!" How can peace exist under those circumstances? It is not possible.

And yet, when Cosmic Intelligence created man in His Workshops, He placed certain seeds within him that were meant to grow into qualities, virtues and talents and, above all, the desire to sacrifice himself for the good of others, all qualities that have been manifested now and then in history by the Saints, Prophets, and Martyrs. They disciplined their lower nature and made it into a channel for the higher Self, rather than let it rule them as it rules the rest of mankind. Yes, but there were so few! Not enough to change the thinking of the masses; and so the few became victims of the majority who did not appreciate them, who turned against them, and who threw them in prison or crucified them. This so terrified the remaining ones that in order not to submit to the same fate, they gave up trying to follow the high example of their predecessors and even turned traitor. What they did not realize is that the law of cause and effect reigns over all our actions. Those who do not follow the way of justice or righteousness are found out by the law sooner or later and, if not massacred like the martyrs, they are attacked by another form of wild beast. Many roads lead to the cemetery! Fear is not the wisest counsellor.

If you will listen carefully, I will explain this situation to you as clearly as possible. Then you will no longer be able to go on behaving as you have. Excuse me for saying so, but at the moment you are so influenced by the actual order in the world that if the Kingdom of God came down on earth today you would do nothing but criticize. "That is not the way to do things, we don't do things that way, it's not right, not normal"! There was once a man who belonged to a tribe of one-eyed people, you know, like the Cyclops. He sailed away on a long voyage during which he landed on an island populated by people with two eyes. When he returned home he told his compatriots, "I met some dreadful monsters with *two* eyes instead of one... we must go and attack those people and do away with their second eye"! That is the way of the world: if you have a faculty, a quality a little more developed than other people, the third eye for instance, if the others find out, they will attack you and get rid of your advantage, saying, "It's abnormal, totally unnatural, we must get rid of it!" The majority sets the standard and, if the majority is brutish, everyone must conform to animal standards. That there might be Angels or Higher Beings in a higher world... no, no, it would not be acceptable. You see what that kind of mentality does? Well, excuse me if I tell you that you think the same way. If I could make you change this way of thinking, your point of view, and make you reason differently, you would see what fantastic results there would be! But would you understand? That is the question.

For years and years I have been talking about the two natures man has in him, the lower animal nature or personality, and the higher Divine Nature which is dormant in most of us because we have not taken the trouble to develop it... the individuality. I have read a great many books and visited nearly every country, I have met many, many people, and I have discovered that no one, not even the most intelligent, the most high-minded, greatest thinkers, know when they are being influenced by the personality and when by the individuality. How could they know the difference when they have no criteria, no values with which to measure? They are convinced that whatever they do, whatever they think, whatever they feel, whatever comes from *them* must be absolute, beyond reproach. If they could see themselves they might also be able to see the diabolical shapes and forms floating around inside them.

When we begin to study ourselves, what do we find? The two natures commingled, entangled, with us in between, not enlightened enough to know which one is influencing us. We do not realize that we must not let ourselves be impressed by the personality... an excellent servant but a very poor master. If we put it to work, if we use it to capacity, it can be very helpful, for it is active, energetic, dynamic, indefatigable and rich! A man who analyses himself thoroughly will understand that he must give his individuality a little more attention, a little more freedom to express itself, to dictate every situation, to command and have the personality obey... in short, to be a divinity. As they are now, humans have extraordinary faculties inside, but they put these faculties at the feet of the personality and do everything to satisfy their lower needs. Show me the exception! There are very few. Man harnesses all his most wonderful qualities, given him by God, to the personality.

There is no doubt that human beings are more intelligent today than at any other moment in history, but what do they do with their intellectual ability, their brilliant minds? They work as spies for their country's Intelligence Bureau, they work at designing more and more deadly arms, they do research in laboratories on lethal chemical products, their higher faculties are in the service of violence and destruction, hatred, war, profit, their time and strength are consecrated to destroying mankind instead of to the realization of the Kingdom of God, to making the world a Paradise for all men to live in. Does anyone do that? Not many, they can be counted on the fingers of one hand. We are not conscious yet, not developed enough to see that it is the egocentric, grasping personality we listen to all the time that is responsible for the abominable state of the world today or, if we do see it, we do nothing about it. I am not the first one to talk about it, but few know, few have ever known what standards, what criteria to apply to every thought, feeling and action in order to be sure that it is right, positive, constructive, in harmony with the Kingdom of God and His Righteousness... or wrong, dedicated to reinforcing the evil forces of Hell, to the destruction of the world instead of its restoration. The disciple must know at every moment of the day and night what is going on inside him, what direction his energies are taking, whether he is being useful and noble, or selfish and vile. If a disciple can learn that, how much more a Master knows, especially as he has all the criteria, all the methods.

Most people, as I have said, are not aware of the two natures that exist side by side within them. Take a pure, lovely, gentle young girl, for instance, and put her down in certain places, under certain circumstances, and you will see what she is capable of, how radically she changes... she can be a fury, a demon! Where did this other totally different side come from? It was always there, asleep. Or take a hardened criminal and set him down in the midst of favourable conditions, you will see what generosity and nobleness of character, what sacrifices he is capable of making to save others... where did that nature come from? It was always there, hidden inside, waiting to be brought to life. People express one or the other of their natures depending on what the conditions are.

To go back to what I was saying: from the moment of their birth, people do their best to conform to the standards and traditions current in the world, in their country, their family, their friends... with no higher ideal than to imitate parents, relatives, neighbours and companions. The lower nature, which wants no more than to live its own life independently, to be free to satisfy its desires and express its emotions with no regard for anyone else, assumes command, and as the lower nature has no respect for divine laws or moral laws, with the pretext of offering us freedom, it urges us to cut ourselves off from the higher world and everything divine. The lower nature refuses all authority, it follows the law of anarchy. Now when a person under that influence comes here to the Fraternité, he brings with him all his old and useless conceptions. The first thing for him to do is to get rid of them. It makes people ill at ease to see the spirit of unity, love and peace, the selflessness that reigns here, the first thing they want to do is to change things... or leave.

Suppose for the moment that all the Masters, all true Initiates decided once and for all to help humans out of their ignorance and selfishness, their chaos, by teaching them to obey the laws of harmony and live in peace, with everyone seeing the truth clearly, everyone being free and independent *within the harmony of the whole.* And suppose an Initiate were on his way to realizing this ideal, with all his disciples participating in this great work in their minds and hearts and wills, all working to form and release intense waves of love in the world powerful enough to awaken the consciousness of all men and to inspire them to form one universal family, instead of fighting each other to the death... what would happen? A person brought up on the old concepts arriving in the Brotherhood at that point would find it completely abnormal, he would feel justified in rebelling against it and criticizing it: "why has everyone such an attitude of respect, why do they spend so much time in silence, why so much concentration on harmony? No, no, it is slavery, hypnotism, sorcery!" He wants to change it all, that is, go back to chaos: in which case the quarrels, the wars, the devastation would be even more destructive than ever, and there would be no end to human despair. The Initiates are trying to put an end to that despair, to develop mankind to the point where man himself will create a better world in which all men are linked and at the same time free. No one loses his freedom, on the contrary, he gains it!

What should we do about those who come here for the first time and criticize us, thinking that what we do is hypnotize people, bewitch or enslave them? Should we go along with their rebellion, back to disorder and conflict, everyone in his corner, isolated, against the world... is that the right path? Or is that the way to permanent dislocation, the very opposite of harmony, brotherly harmony?

I should like to ask the young people of the world this question. Suppose you did succeed in destroying things as they are now in the world and you had no one to obey, no laws, no leaders, no moral authority, even though people do exist in the world whose pure lives and higher knowledge give them genuine authority (but you prefer no authority at all, you prefer Anarchy), and suppose you destroy society as it is... how will you live? You will go back to solitude and isolation, to killing each other, and then you will have to restore the same social laws, the same customs you destroyed! Besides which, you will have chosen a leader born in Anarchy, who will be more tyrannical than anyone and, having created him, you will have to submit to him and his ideas.

Yes, that is the point: the same institutions you so dislike will have to be built up once again, you will have to reinvent the whole social organisation, the only way man has found so far of getting on together, sharing the work in order to exist. As it is, the butcher and baker, shoemaker, lawyer, soldier, all work for everyone, not only for themselves; train conductors work for all who want to travel, doctors heal the sick, other people make clothes for you! You too can do something with your life to benefit the world. Each one in this society does something that facilitates the life of everyone else. Tear down that order and you have to begin all over again, you will have to spend your days looking for food like an animal. It will be all over with culture, all over for the arts and the rest of civilisation... man will go back to hunting and fishing and scrounging for survival. At least now you have an occupation, thousands of people have occupations, they contribute their work and you contribute yours. Throughout the ages man has always found it easier in the end to live in society.

And that is not all. It is only in appearance that men have solved the problem of the collective life: if outwardly they are part of society, inwardly they remain isolated, aggressive, hostile to each other. They are at the same stage as troglodytes, each one separate in his hole. That is why I insist on inner development, the opening up of their innermost heart in order to unite with the forces that will bring universal brotherhood on earth. Men believe themselves linked, but inwardly they live their own lives quite separately, like troglodytes. If you analyze yourselves, you will see that I am right.

What we are doing here is to bring a higher degree of consciousness to the world, we are the example here of a new society in which men are free and happy, first of all within themselves, and then outside for all to see. And so people who come here bringing their old ideas with them, are absolutely useless, they do more harm than good. Let them go somewhere else and suffer a little in order to learn what it is like to live alone with no one to help... that will bring them wisdom! For the moment they are far from it.

To understand what it is we do here, let us go back to the sentence, "During the silence the disciple should participate in the work of the Master." What is the Master's work? To steal your money, separate the members of your family, rob you of your house, your car, your possessions? No, dear brothers and sisters, the work of a Master... but you cannot imagine what a Master does, you are too involved with your schemes and worries. A Master is free of all the things that absorb you, he has no personal worries, his thoughts are all trained on subjects you cannot imagine, he works ceaselessly for the good of the whole world, not for you only, nor for himself. You are engrossed in your affairs, your daily life; a Master is detached, disengaged from everything on that level, he has no wife, no mistress, no children, no obligation nor impediments, and he has solved his problems... his mind is entirely free for more grandiose projects. Even if you have been brought up in the old tradition it is worth your while to help him if only with your thoughts. Say to yourself, "Poor fellow, look at him doing this gigantic work all by himself with no help, no encouragement, no reward... I will give him a hand." But you would rather criticize him. Do you know what criticism does to him? It stings and cuts him, it hurts him... that is what criticism does. Criticism is a wound to a Master.

Have you any idea what a thought is, what your thinking can do? It can hurt worse than the sting of a wasp, and it can be as healing as the touch of an Angel! People emit no matter what thought, a critical one concerning someone who is doing his best to help them, for instance... thoughts that tear him apart. They have no idea they are doing it, but if they go on that way, the being they are thus attacking will close himself off from them, withdraw his aid and leave them to grope round in the dark by themselves.

I know that those who come here for the first time will contribute nothing but criticism, doubt, misgivings and in-comprehension, because they are not prepared. But I also know that in time they will change, they will see how wrong their reasoning was and they will adapt to our ways, where-upon they will open themselves, blossom out, and beam! Nothing will hold them back. But that will not happen at once and in the meanwhile I must bear all the negative thoughts they send me, accept the bricks they throw without complaining or becoming discouraged, I must even look lov-ingly at them in spite of everything they hurl in my direction. Sometimes they understand what a tremendous task I am try-ing to accomplish and they decide to help me. And those are the very ones who advance in leaps and bounds... they go further than anyone! There was no one more dedicated to the slaughter of Christians than St. Paul, but once he received his lesson on the road to Damascus, he began to spread Christ's teaching with more fervour than anyone! That is the reason for my patience, I am waiting. And the wonderful thing is that one day, those will be the ones I will be able to count on.

The great Universal White Brotherhood has existed since

the beginning of time. It is composed of Saints, Prophets, Initiates and great Masters, Angels, Archangels, Divinities. Every now and then these older Brothers decide to send the world a Saviour; we have had Ram, Buddha, Lao-Tse, Fo-Hi, Pythagoras, Hermes Trismegistus, Moses, Jesus, Peter Deunov... all bringing the eternal principles to man in a *new* form. It is always the same Teaching, for Truth does not change, but, as mankind evolves, the methods and means, the *form* must change, it must not remain the same forever. As your child grows bigger, you do not force him into the same baby clothes, you know he needs a bigger size and you buy him new clothes. So it is for religion and philosophy, each era requires a new form.

Now an epoch is coming when the Universal White Brotherhood above will manifest itself here below. Jesus prayed to his Father, "Thy Will be done on earth as it is in Heaven", which means, As above, so below. The Universal White Brotherhood above directs the universe; this little Brotherhood on earth is a reflection of the great Brotherhood above. I have never considered the Fraternité as the real Universal White Brotherhood... I am more realistic than people think, I do not deceive myself... but one day it will be the same. If we can make ourselves into a temple to receive the Older Brothers, then our little Brotherhood will become one with the August Brotherhood above. That is the way I think. My work, my task, consists in helping the Older Brothers by preparing the way for them to come and manifest themselves through the thousands of purified souls that are waiting for them. When that time comes, the world will sing, it will be an end of war, of frontiers between the nations, and happiness will reign. Is that task not a worthy one? Don't you want to work that way also?

Above all, I ask you not to try to convert me to your way of thinking, your philosophy. I know all about that, I have been through it. What you must realize is that I know things you do not know: accept that, it would be so much easier for you. If not, there will be others who will take your place, thousands are preparing themselves to serve me... and where will you be then? When you realize you will shed a big tear and lament, "Ah, how stupid I was! I lost an invaluable opportunity. Woe is me!" But the opportunity, once you have let it go by, cannot be recaptured. You can pray all you like, there will be no answer.

Now, back to the personality. There are so many different forms it takes! In most women, the personality manifests itself as possessiveness, the desire to hold, to contain, to keep for herself. In men the personality manifests as the desire to conquer, to govern. Women want to possess, and this desire is accompanied by jealousy. If I receive a sister for a few minutes more than another sister (because it is expedient) the second one, instead of rejoicing at her sister's good fortune, flies into a rage! This fault must not be allowed to last a whole lifetime. Jealousy is worthy of your attention.

Everyone who comes here is put to little tests from time to time, to find out what they are made of, how much is inside them. I take a pin, symbolically, and jab it into someone; if he reacts against it, I think to myself that his personality is very much in evidence and, depending on what he says, I know where to classify him and what to do for him. Everyone is put through these little tests without knowing it, only later do they understand. It is not to hurt anyone, of course, it is in order for them to know themselves. No one knows himself, people have no idea who they are and what they are like... they are convinced of their perfection, convinced that their judgment is divine. How do I convince them that they do *not* see clearly, that their vision is all too human? By putting them through certain tests that will show them clearly what they are like and destroy certain illusions they have about themselves. Know thyself!

We should rejoice over our neighbour's good fortune... a quality that is hard to find. Here is a terrible story on the subject. There was once a king, who, wishing to reward one of his ministers for services rendered, said to him, "You may ask for anything you wish, you shall have it." But, knowing how much the minister envied one of the other ministers he added, "But whatever you ask for I will give double the amount to so and so." The minister pondered. Finally he said, "Your Majesty, I ask that your soldiers put out one of my eyes." His enemy would thus lose both his eyes and be blind. It is a terrible example, but it is human nature, humans are like that. Instead of rejoicing over someone's good fortune, they want to know, "Why not *me* ?" If you give a slightly bigger apple to one child, all the others will be incensed! Ah! Human nature is really something. The individuality, on the contrary, wants nothing more than the happiness of other people... like a mother and her child, the Divine Nature is very clearly expressed in a mother's sacrifice, she goes without food for his sake, she stays up all night nursing him... that is the individuality.

My sole desire is for you to have the best, only the best. If this were not true, long ago I would have gone off taking everything I could get from you, saying, "Bye bye! See you next time!" Such opportunities have not been lacking, several immensely rich women have asked me to marry them and abandon the Brotherhood... Heaven knows why. There are witnesses to that. But I never accepted. And I never made a fortune! But I have been free to help you... otherwise where would I be now and in what condition? And so, if I have not robbed or cheated you, what is there to complain about?

The thing that counts most for me is your wellbeing, I want to see you full of light, levelheaded, clear and transparent, powerful, happy and strong. Why? Well, don't tell anyone, but for my own personal satisfaction! I want to be able to say to myself, "You succeeded after all, old fellow!" Who would then be speaking, the personality or the individuality? I leave the answer to you. One day I told you that there was no such thing as a completely disinterested action. Even God has an interest where we are concerned. Perhaps the greatest saints were also the most selfish people, the good they did was for a reward: a seat next to God! What about a mother... is she entirely disinterested or is she thinking about her old age: "My son, my daughter will take care of me." Disinterestedness is nothing but the smallest degree of interest! And what is interest but the smallest degree of disinterestedness! Hatred is the smallest, the least degree of love... and what is love but....

Actually there is no such thing as complete disinterest, complete unselfishness, it is always some degree of interest or selfishness. Even when all you want is to make men happy, luminous and free, it is still a form of self-interest: you want to become divine. But this interest is so completely disinterested that it no longer belongs in a human category, it is divine! Actually, one always has some selfish interest, some self-concern, some advantage to be gained. My consuming interest is to leave a trace within each one of you of the Divine World, ineffaceable traces so that later when you are far away from me, you will remember. A selfish interest?

Le Bonfin, 11 July, 1973

Chapter 22

Toward Universal Brotherhood

II

"In the universe everything sings together in unison and perfect harmony and, in this Cosmic harmony, each creature has a note of his own, a special intonation. To be in tune with the universal harmony man must know his own tone, and which music to play. The higher he goes, the more harmonious and pure the sounds will be."

There, dear brothers and sisters, is a very rich thought. Everything in the universe should sing in harmony with everything else! Man has gone so far from the truth he has lost his ability to vibrate in unison with the Cosmos. That is why no one thinks of the universe as one tremendous symphony. This subject has many interesting aspects, which I will try to make clear to you.

First of all, this idea of a symphony... if you were at home alone, you would be able to sing or play the instrument of your choice without having to think about being in tune with anyone else, only with yourself, your own mood, your own state of mind. If you are happy you will sing a gay tune, and if you are sad you will sing a sad one, no one else matters, only you. That is the separate, isolated life. Many prefer this life to any other, they would rather sing, work, think and act alone than be in tune with something outside themselves, something greater, something Divine.

Why do the great Masters advocate living in a collectivity? Because it brings about a change of consciousness. Instead of behaving in an independent, disharmonious way, the disciple must become attuned to other people; it is this synchronization with the group that makes him progress, he gains tremendously from trying to live harmoniously with others and lead a more reasonable, luminous life. The group is striving to be in harmony with the larger Cosmic collectivity, and each member strives to enter into communication with Cosmic Intelligence, with the result that he is showered with blessings! The opposite is also true. A man who is against others, against brotherhood (which is the symbol of Cosmic Brotherhood) does himself serious harm and is apt to fall ill as a result, or even do himself in. There can be no progress, he will not advance or evolve as long as he is in disharmony. To believe, as so many people do, that to be constantly in opposition, rebellious and anarchistic, demonstrates how intelligent, powerful and deserving of happiness one is... is the greatest of errors. Where does the idea come from, why do we go on believing it? Initiatic Science has never said that man can improve himself or obtain the blessings and riches of the universe by living separately, in Anarchy. What is the reason for this tendency so prevalent all over the world at this particular time in history? For me the reason is ignorance, the abysmal ignorance of mankind.

The first thing to learn is that by being in harmony with these higher Laws, Forces, Intelligences and Powers that far surpass us, we improve first of all our health, and then other things little by little. The fact that the average way of thinking and acting does not show dire results at once leads people astray, they do not see the good or the bad effects of their action. When they are in a state of chaos, there are no immediate reactions, they are the same, even sometimes better!

Humans are fooled by that. As I have already explained, if Cosmic Intelligence arranged things this way, it is to give us time to change, to acquire wisdom. The Love of God, instead of punishing us immediately, gives us time to make repairs. In the world when a man transgresses the law by falsifying accounts or records he is not caught for months, sometimes even years. It takes a long time for the authorities to verify the records and during that time, long before the day of judgment, the "treasurer" in question has time to repair his transgression.

God also gives man a chance to correct and rectify his errors... it is one aspect of His Love. It fools us nonetheless, for as our punishment is not immediate we are apt to think everything is all right and we congratulate ourselves on having been successful... the perfect crime! We don't realize that everything is recorded.

In essence the page I read you says that man must now change his state of consciousness. Instead of continuing with his ancient concepts, never questioning whether his behaviour might be less perfect than he thinks (perhaps even harmful to some of the visible and invisible creatures around him), he might try asking himself how to be a help instead of a hindrance, how to win the approval of the Higher Beings. When people's vision is restricted to their own personal lives and narrow field of consciousness, the capacity to feel and to create becomes more and more limited until finally they are too weak and too worn to be of any use, others who are stronger trample on them and they land on a heap of fertilizer! The whole world, especially the young, have adopted today's deplorable philosophy with unbelievable enthusiasm. They are destructive and violent because it suits them to be that way now, but sooner or later they will be ground to bits by Cosmic Law, if not by human laws.

The more seriously a disciple takes into consideration the philosophy of the Initiates, the more he tries to lead a life of love and gratitude in accord with that philosophy, and controls his thoughts and attitude, the more his consciousness will be enlightened and enlarged. The Divine Life itself will see to it. He will experience, see and feel things he never dreamed existed, and overflow with gratitude toward Heaven for his new consciousness. One thing must not be overlooked: this enlarged consciousness must be accompanied by change, it must be part of a tremendous change in him. Instead of devoting his energies to the personality, he must now devote them to the individuality. Instead of wondering all the time how to satisfy his ego, he must submit to the other centre... harmony, beauty, light, Infinity, Eternity... God! He must strip himself of all burdens and all obscurity, he must stand upright and advance into the new life beginning to circulate in him. How simple, how clear and easy!

"Not so easy," you say. Of course nothing is easy if your centre is the personality. But if you change your centre and your goal, the point of convergence of your energies, it is surprising how easy everything becomes! Years ago I spoke on the subject of the three systems, egocentric, biocentric, and theocentric. You may not be familiar with the three systems, but you know what a goal is, and you know there are means to a goal. Well, the goal man has chosen should be the means, not the goal, and the means he now uses for his own ends should be the goal... what does that mean? It means that men have the goal of contenting the personality, of satisfying their appetites, their instincts, their ego. That is their goal. And the means? All their innate intelligence, knowledge, ability, are used to satisfy their needs. Yes, the means which are divine are used to obtain a diabolic goal. It should be the reverse. The goal must become the means, the physical body, stomach, sex and impulses of the lower nature should be the means to reach the divine world, the Light. If you can make this change in your way of thinking, your life will be transformed.

Unfortunately, people think that science, philosophy, art and Heaven itself are there to cater to their prosaic appetites and interests. Where is it written that man should devote his life to those things? We must take everything we have been given by God as the means to reach the divine Ideal. Then it becomes easy, those who have done it agree! It is up to you... if you have doubts, if you continue to hold on to the old concepts inherited from the past, you will never free yourself or be really healthy, really happy. You think of the Initiates as eccentrics, lunatic or sunstruck as the case may be, and you prefer to follow the rabble, ignorant as it is, because it is the majority. What do you gain? You can always find a few crumbs to nibble on, but then come the inevitable hard knocks, the blows, the suffering. That is the usual life, a little joy now and then, a little pleasure for which you pay very dearly. If you listen to the Initiates and say to yourself, "Today I am going to start putting all the means God has given me into the service of the High Ideal, the great Universal White Brotherhood, so that the Kingdom of God may be installed on earth", you then come under a new order, you enter another dimension and join a group of highly intelligent, advanced creatures who see your decision and come to inspire you and help you. From then on you go from one marvellous experience to another! Why not believe this?

You are still clinging to the old philosophy, I can see it. You think, "This is nothing but talk!" But those who choose not to understand will come a cropper eventually... it is the law. Alas, only when it is too late do we realize we have taken a wrong turn. Young people say, "Leave us alone, let us have our own experiences, then we will understand... as you did!" A grandmother was telling her little granddaughters how to behave, how to be chaste and virginal. "And you, Grannie," asked the little girls. "Did you have no experiences at our age?" "Yes, yes, but then I learned to behave."

"Well, let us have a taste of life, like you, and then we too will behave!" This is only normal.

If you wait too long before starting to behave wisely, people will not believe in your wisdom, it is when you are young that you should set the example. As you see, it is already too late for me, I am old! But even when I was young, I thought this way and said these things... the trouble is, when you are young no one believes you, they wait until you are old! Then, yes, they are more apt to be convinced. No one believes young people. In order to be heard you must grow old, as I did!

But let us leave all that and go back to the essential. Every one of us can be compared to a musical instrument, a clarinet, trumpet, violin, piano, guitar... and the divine Life blows through the instrument or plays on its strings. Each one of us has a determined sound, Cosmic Intelligence has attuned us all so that we form a universal symphony, but not here, the symphony does not exist on earth. Because of their instincts and emotions, men do not vibrate, as God meant them to, in unison with the universal harmony. Their consciousness is too limited, they are too concerned with the personality and its desires. The day they begin to understand this idea of collectivity, or rather of brotherhood, and take that idea as the single great goal of their lives, then they will vibrate in tune with the universe and because of that accord they will be able to receive the beneficent currents from the Cosmos, their channels will be open to the celestial energies. A strictly personal life blocks those channels and energies... a fact that will one day be known all over the world, children will be born knowing that nothing is worse than a selfish life, for it is out of tune with the Divine Life, it is the beginning of all human troubles and unhappiness.

We must return to the original harmony created by God. We were created to be in tune with each other, to be part of the Cosmic orchestra, but we have never really understood what an orchestra, a chorale, is. Our physical body, when it is in perfect condition, is a chorale in which all the cells and organs sing together, producing joy, health, a sense of wellbeing to the entire system. Cells and organs that do not sing in harmony cause the human system to fall ill with the illness produced by the discordant sounds it emits.

We have never interpreted the fact that a musician playing in an orchestra has no right to play whatever comes into his head... he must confine himself to the notes and tempo of the particular music the rest of the orchestra is playing, otherwise he will no longer be part of that orchestra! At the moment believe me, mankind is not a good orchestra, all you hear is dissonance, a cacophony of sounds that offends the ears! Everyone thinks he has the right to sing or play what he wants. In an Initiatic School you learn true harmony. First of all, to be in harmony, you have to understand that harmony is preferable, and then you must long for it with love and finally you must decide to make the effort and sacrifice which will bring it about. And then? Then there will be no need to keep trying, harmony is there, it can be felt, it speaks for itself. That is the reason the Initiates say, "Be silent."

I have talked to you before on this subject of the Initiatic formula, "To know, to will, to dare, and to be silent." Once you learn, once you *know* what is good, what is right, and *want* that at any expense, once you *dare* to launch out into the great work, then there is no need to speak, the good speaks for you, expressing itself through you whenever you are *silent.* When you are joyful and happy, do you have to tell everyone, or can they see it? And if you say, "How happy I am, what peace, what harmony I feel!" when actually a revolution is taking place inside you, no one will believe you, for obviously you are unhappy, the inner scarecrow is clearly visible! Do lovers have to say they love each other and are happy together? They contemplate each other in silence, confident that the other feels the same harmony, the same joy. It is only when they no longer love each other that words are necessary... to hide their true feelings.

Words often do nothing but spread confusion, whereas silence always expresses the truth. You say, "Ah, I see. You talk so much in order to fool us, to deceive us!" Possibly, possibly, but I did not say that words are *always* deceptive, I said that words are used sometimes to hide the truth... which is true. There is no need to talk about what exists, truth is self-evident. It is when nothing is going right that we must lie, to deceive others into thinking things are all right. When a boy no longer loves a girl, that is when he feels the need to tell her how much he loves her, and she, poor thing, doesn't realize that the words are taking the place of feelings that are absent! Others will try to hide the intensity of their love for someone, but they cannot, the feeling is transparent. Even if they never declare their love, even if they say nothing about the way they feel to the object of their love, nevertheless it is a great burning fire apparent to all! And what about hatred, can you hide the fact that you hate someone? It is the same as love, it transpires.

Forget everything else I have told you if you will, but remember this: to lead a life which is entirely personal and selfish is the way to unhappiness, and to lead a life for the collectivity, thinking about others and working for universal brotherhood, is the way to joy and every blessing. I believe that. Only, as I know how slow the laws are to act, I force myself to be patient. I say to myself that to be conscious of the collectivity, to be aware of the whole, is the first blessing. The Universal White Brotherhood will be realized here on earth, thanks to those who work for it. Those who are still alive will see it realized, and Heaven and earth will vibrate differently for them. The trouble is, where is the faith that makes one believe and decide to do something about that belief? We are lacking in that kind of faith. People wait to *see* before they allow themselves to believe, but then it is too late. The laws move slowly, but they are absolute. I do not need to read books on philosophy or argue with learned people in order to be convinced of something, I verify for myself, and then no opinion can convince me or dissuade me... I verify for myself and, if people do not believe me or agree with what I have found, then what can I say but, "Well, poor thing, since you are so hard-headed, since you refuse to be convinced, what will happen is that I will go on receiving blessings and you will go on being miserable!"

It is the slowness, the delay in the proceedings that keep us from understanding. In the past, the Initiates who discovered the secret of the Elixir of Eternal Life could prolong their lives for centuries if they chose, observing during this time the way the laws work. They were able to witness the rise and fall of great families who caused their own decline and fall and eventual disappearance, by having fought their way to power and supremacy with criminal methods. They were able to observe individuals who, although disgraced and dishonoured, held to the narrow path despite all and became magnificent divinities... one must live a long, long time in order to verify the laws.

The collective life, the unlimited, fraternal, universal life, is the beginning of progress, of all unfolding and blossoming. And the personal life, that is, selfish, narrow, obscure and miserable, is the beginning of limitations and obstacles for all who are stupid enough to live that way, like insects. Why have insects not evolved over the last few million years? Because they chose the philosophy of the personality! Microbes

have gone even further down than insects. Whereas the Angels, Archangels and Divinities chose the path of brotherhood, of universality, and they advanced and reached upward
further and further until they became one with God!

A man who chooses the way of brotherhood is a wise man.
The others grow weaker and weaker until they disappear.
That is the law. They may live, but they do not live the glori-
ous life.

These are laws I have verified, dear brothers and sisters,
and now, even if I am offered all the riches in the world in exchange for my philosophy, I will refuse, I will hold on to this
philosophy because I know what it will bring me: Heaven
and earth. Why would I want to go along with the deplorable
philosophy of others, the philosophy of the personality? "I
want to live my own life!" they say. Well, that is exactly what
they will do, with tears and grinding of teeth! If they said, "I
want to consecrate my life to the High Ideal", it would be the
beginning of the real life. Consecrating one's life to something
great and glorious does not mean that you die, it means that
you change the goal of your life, your destination. Life then
becomes indescribably beautiful and poetic. "I want to live
my own life!" Yes, a stupid, foolish, useless life! But that is
man's reasoning, especially the young. When I hear someone
say, "I want to live my own life", I know in advance down to
the last detail what that life will be full of: trouble, difficulty,
sorrow.

Now I will explain what it means to play one's own violin.
You will see that not many musicians know what a violin
really is. It is the symbol of man. It has four strings that correspond to man's heart, mind, soul and Spirit, and a bow which
is his will. A virtuoso tunes his violin before beginning to
play... why? We never think of tuning our violins in life, and
that is why we do not play very well and draw only scratchy,

grinding noises from it! We complain, "I don't know what's wrong with me!" We forgot to tune the violin. In tune with what? A diapason exists for that very purpose, but people prefer to remain ignorant; it makes them suffer and that is what humans like, to suffer.

I have been given the tremendous privilege of hearing the Music of the Spheres, Divine Grace accorded me this great honour once when I was out of my body, in the Spirit, and I know now that creatures and trees, mountains and stars, sing in unison with Creation, I have heard them. All creatures sing together in harmony, only this Music is imperceptible to the human ear. Perhaps one day after much effort on your part, you will be allowed to hear the divine concert of life in Nature.

Keep this thought in mind, that you must do everything you can to extract yourself from the grip of the personality and expand your thinking. Accept the great Universal White Brotherhood as a point of departure for your advancement, your evolution: it is the only way. If happiness and abundance are not yet triumphant in the world, it is because mankind is still divided, everyone is watching out for his own good and no one tries to break out of the narrow circle of the personality. With those conditions in existence, how can the Kingdom of God come into the world? The name Universal White Brotherhood presupposes another way of thinking and living, with other methods and another goal: the Kingdom of God and happiness for all mankind.

Once I interpret the three words Universal White Brotherhood for you, you will understand how you should work, you will understand why. The three words contain everything, it would take me years to explain everything they mean. The name bothers some people, it seems they do not like it. They will, once they know what it means. It means we must enlarge our consciousness, our thinking, to the extent of including the whole world, with no other goal in our minds than that of contributing to the good of the world. Because good is collective, all who belong to the collectivity are sure to benefit. A collectivity that is in trouble brings trouble down on everyone in it.

People have no idea how to behave, each one works for his own goal, which means that nothing is sure, if some misfortune such as world war for instance, hits the collectivity, each member is bound to be affected adversely. As long as the collectivity is exposed, each individual is also without shelter, but when everything goes well for the collectivity, then everyone is protected, everyone helps everyone else! The thing that is stupid about working only for oneself is that it is so costly, there can be no end to war, devastation, hunger and suffering. The Third World War is on its way now to teach humans a lesson they refuse to learn any other way. Each one still works for himself, his own interests *against* the interests of the community, each one pulls the covers over to his side, imagining that he can enjoy security regardless of what happens to the others. Well, no. If something terrible happens to the collectivity, then each man's personal happiness will go by the board, as part of the collectivity he is vulnerable. How can people fail to understand what is so obvious?

Imagine that you play an instrument, you are part of an orchestra. All you do is play your score, your part, while others play theirs, yet you are surrounded by wonderful sounds that delight your soul, all the harmony and beauty coming from the whole orchestra is yours simply because you are part of the whole! And the same for a chorale, all you do is sing a few notes now and then, and you are bombarded from all sides by harmonious, poetic music and voices that fill you with joy.

What we must do from now on is work together in harmony, to improve the collective life in order for all men to be happy... forever. As long as each individual remains fastened to his own private property, there will always be a sword of

Damocles hanging over his head threatening him, because he is part of a whole that is not properly oriented. Why not understand that? If we did, it would bring about unity, and that in turn would bring happiness to the whole world.

May Light and Peace be with you!

Le Bonfin, 16 September, 1973

The Complete Works of Omraam Mikhaël Aïvanhov

Translated from the French

PRINTED IN FRANCE
FEBRUARY 1985
PROSVETA EDITIONS, FRÉJUS

– N° d'impression : 1395 –
Dépôt légal : Février 1985
Printed in France